Monographs on Astronomical Subjects: 4

General Editor, A. J. Meadows, D.Phil.,

Professor of Astronomy, University of Leicester

Applications of Early Astronomical Records

In the same series published by Oxford University Press, New York

3 Cosmology and Geophysics
P. S. Wesson

5 The Nature and Origin of Meteorites
D. W. Sears

Applications of Early Astronomical Records

F. Richard Stephenson, Ph.D.
Institute of Lunar and Planetary Sciences,
University of Newcastle upon Tyne

David H. Clark, Ph.D.
Royal Greenwich Observatory,
Herstmonceux Castle, East Sussex

Monographs on Astronomical Subjects: 4

OXFORD UNIVERSITY PRESS
NEW YORK

First published in the United Kingdom 1978 by Adam Hilger Ltd., Bristol.

Adam Hilger is now owned by The Institute of Physics.

First published in the United States 1978 by Oxford University Press, New York.

ISBN 0 19 520122 1

Filmset by The Universities Press (Belfast) Ltd, Belfast, N. Ireland and printed in Great Britain by The Pitman Press, Lower Bristol Road, Bath BA2 3BL.

To Gillian, Susan, John,
Matthew, Andrew and Stephen

Contents

Preface

Our objective in writing this monograph is to discuss the principal ways in which pre-telescopic observations can be used in present-day astronomy. Both of us have specialised in this area of research for many years, and we have found it a fascinating study.

Observations made with the unaided eye are undoubtedly of low precision, but there are several fields in modern astronomy where this early data is of unique value. Thus it is possible to study the variation in the rotation of the Earth with reasonable accuracy over the last three millennia. Again, not a single supernova has been seen in our own Galaxy since the telescope was first used in astronomy, but anything from six to eight galactic supernovae are recorded in history between AD 185 and 1604. As a further example, regular sightings of sunspots were recorded in Far Eastern history at least as far back as AD 300. These applications are those for which we have special expertise or interest, and a chapter is devoted to each. Other possible applications of the historical records are mentioned in chapter 1.

We feel that present-day astronomers — professional or amateur — should be proud of their heritage. In no science other than astronomy are the written records of the past so directly relevant to current research.

May 1977

F. Richard Stephenson
David H. Clark

1. *Sources of Historical Astronomical Records*

1.1. Introduction

With the invention of the telescope and its application to the study of the heavens, the age of modern astronomy can be said to begin. Prior to this time, no matter how skilful the observer or how well designed his instruments, only a very limited accuracy of observation could be achieved.

However, the time span of recorded pre-telescopic astronomical observations is long, roughly an order of magnitude greater than that of telescopic observations. Regular astronomical observation goes back at least as far as 700 BC. Astronomers would thus be ill advised to underestimate the vast body of ancient observations which has come down to us.

In this chapter, we give examples of the various kinds of astronomical observation recorded in history before Galileo, covering as wide a range as possible. The remaining three chapters are devoted to an investigation and analysis of selected types of observation. These are among the most important in present-day astronomical research.

Only a few civilisations throughout the world have contributed significantly to the records of astronomical phenomena from the historical past. Pre-eminent among these must be Europe, China (together with Korea and Japan), Babylon and the Arab World. Data from these named sources are readily accessible. As a result, we have decided to confine our attention to them. Another limiting factor in our choice is our own specialisation: we do not profess any expertise in the astronomy of India or the Mayas, for example.

In what follows we have preferred to let the records speak

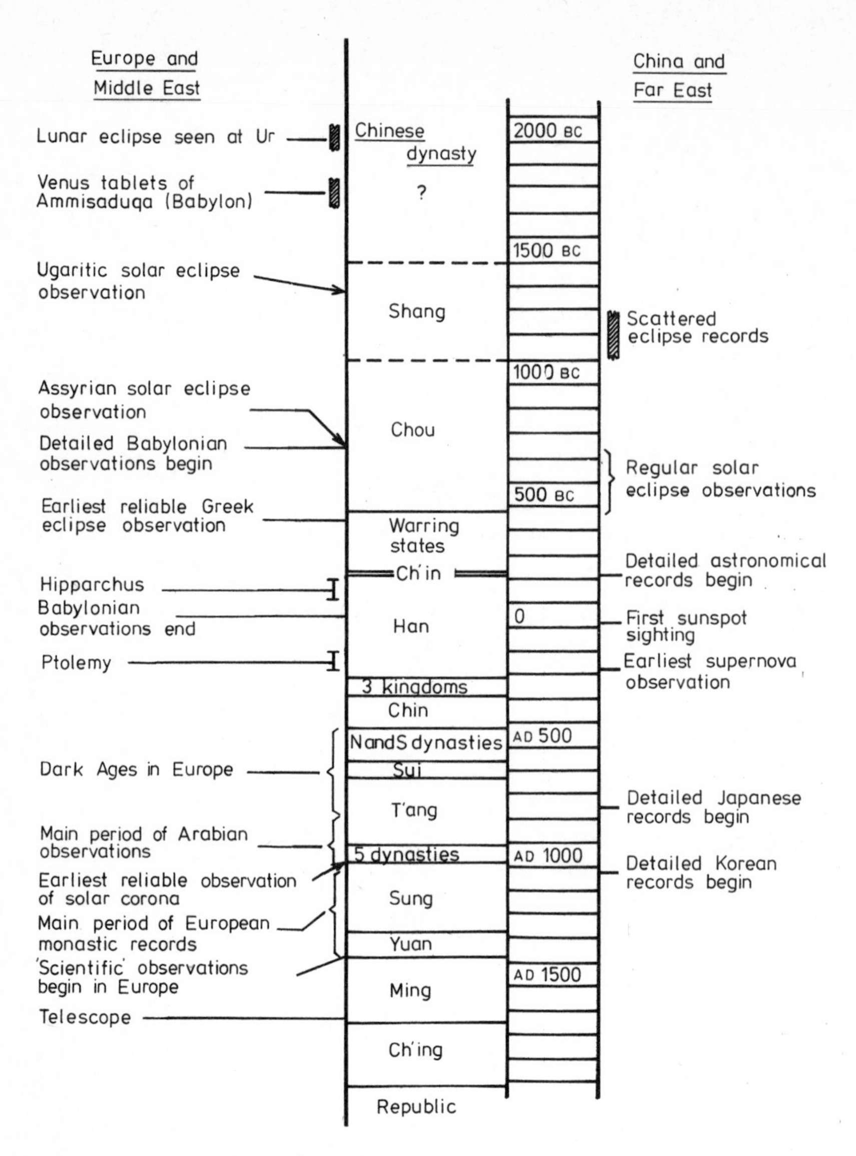

Figure 1.1. Highlights in pre-telescopic astronomy.

for themselves, by selecting examples from the various types of observation available. Figure 1.1 illustrates the parallel development of astronomy in the selected regions.

1.2. Ancient Europe (before AD 500)

Astronomical observations from Europe in this period are essentially confined to two principal sources, (i) the Greek and Roman Classics and (ii) Ptolemy's *Almagest.* In both sources only a small fraction of the original observations has survived, and it is difficult to form an assessment of the extent of these observations.

The Classics are mainly historical, philosophical or poetic works, and by and large only the more spectacular astronomical events (usually eclipses and comets) are alluded to, normally in very general terms. Because of the nature of this material, precise dates and places of observation are seldom given. Attempts to deduce the dates of some of the large solar eclipses mentioned in the Classics were a preoccupation with astronomers around the beginning of the present century, but their efforts met with very little success. At the present time, many of the dates are still in doubt. A good example is the so-called 'Eclipse of Hipparchus.' The fact that Hipparchus of Rhodes towards the end of the second century BC used a solar eclipse in his determination of the Moon's distance is noted by Cleomedes, Pappus and Ptolemy, of the early writers whose works have survived. The eclipse was observed to be total around the Hellespont and was estimated to be $\frac{4}{5}$ total at Alexandria. Unfortunately, no-one troubles to state whether it occurred in Hipparchus' own lifetime, or whether it was an earlier (possibly much earlier) event.

Between the foundation of Alexandria (331 BC) and the death of Hipparchus (*c* 120 BC) there were three total eclipses which reached a very large magnitude at the Hellespont. The dates of these are BC 310 August 15, BC 190 March 14 and BC 129 November 20. All would give about the right magnitude at Alexandria, if total at the Hellespont. Muller and Stephenson (1975) had some preference for the later date, but Toomer (1974) concluded that the date was 190 BC. The issue is still very much open to debate.

In his all-embracing *Almagest,* a compendium of Greek astronomy, the great second-century Greek astronomer, Claudius Ptolemy of Alexandria, collected together a substantial number of observations of various kinds going back several centuries. These included lunar eclipses, equinoxes and solstices, occultations and close approaches of the Moon to stars, and close conjunctions of planets with stars. In

Ptolemy's treatise, most observations cited are used as examples to illustrate various aspects of his orbital theory. Many of these observations are of considerable interest today, but, undoubtedly, Ptolemy passes over in silence a large body of similar data which would also have been of value. Still, we must not be too critical of Ptolemy, for without his efforts virtually all of the observations which he quotes would have disappeared.

Examples of the various types of observation found in the above sources are as follows. (Translation of the majority of the classical works has been performed by one of the authors (FRS) unless a specific reference is given.)

1.2.1. Solar Eclipses

Thucydides, II, 28. 'During the same summer at the beginning of a lunar month (the only time it seems when such an occurrence is possible) the Sun was eclipsed after midday; it assumed the shape of a crescent and became full again, and during the eclipse some stars became visible' (translated by Smith 1919).

This is the earliest reliable European observation of a solar eclipse. Thucydides is writing as a contemporary in his account of the Peloponnesian War. There can be no question that the eclipse is that of BC 431 August 3, which was annular. As Thucydides does not mention the central phase, we must presume that the eclipse was only partial where it was seen. The place of observation (Athens or Thrace?) has been hotly disputed, and Thucydides statement that more than one star was visible is also doubtful (only Venus should have been seen).

Marino Neapolitano conc Proclus, XXXVII. 'Nor were there portents wanting in the year preceding his death; for example, such a great eclipse of the Sun that night seemed to fall by day. This happened in the eastern sky when the Sun dwelt in Capricorn . . . .'

The Greek philosopher Proclus died in AD 485, so that there is no difficulty in fixing the date of the above eclipse as AD 484 January 14. When this occurred the Sun was indeed in Capricorn, and, as it happened near sunrise in Greece, it would be 'in the eastern sky.' This is probably the most reliable of all solar eclipses reported in the Classics. It is only a pity that there is no precise mention of totality.

1.2.2. *Lunar Eclipse*

Ptolemy: Almagest, VI, 5. 'The seventh year, therefore, of Ptolemy Philometor, which is the 574th of the era of Nabonassar, from the beginning of the eighth hour until the end of the tenth, on the 27th or 28th day of the Egyptian month Phamenoth, was seen at Alexandria the Moon eclipsed to the extent of seven digits, from the northern limb. The middle (or half-way through) of the eclipse coincided with $2\frac{1}{2}$ seasonal hours after midnight.'

The beginning of the era of the Babylonian King Nabonassar corresponds to 747 BC, and the date of the lunar eclipse is equivalent to BC 174 April 30, the occasion of a partial lunar eclipse visible in Alexandria. At this period, times were measured to no better than the nearest third of an hour. The estimate of the magnitude of the eclipse is to the nearest twelfth of the Moon's apparent diameter, in other words the eclipse was observed to be a little more than half.

1.2.3. *Comet*

Seneca: Naturales Quaestiones, VII, 15. 'After the death of Demetrius, the King of Syria, whose sons were Demetrius and Antiochus, a little before the Achaean War, a comet appeared as large as the Sun. At first it was a fiery red disc, emitting light so bright that it dispelled the night. Then gradually its size diminished and its brightness faded. Finally it disappeared entirely' (translated by Corcoran 1972).

Demetrius Soter died in 151 BC, which fixes the date. However, among the Far Eastern observations of comets there is no obvious candidate for the comet to which Seneca refers. Possibly, as Seneca wrote long after the event (some two centuries), he is giving us an inaccurate account.

1.2.4. *Occultation*

Aristotle: De Caelo, II, 12. 'For the Sun and Moon perform simpler motions than some of the planets, although the planets are farther from the centre and nearer the primary body, as has in certain cases actually been seen. For instance we have seen the Moon when half-full approach the planet Mars, which has then disappeared behind the dark edge of the Moon and reappeared on the bright side. Similar observations about the other planets are recorded by the Egyptians and the Babylonians, who have watched the stars from the remotest past....'

The above is one of the very few astronomical observations which Aristotle himself is known to have made. He only cites it in view of its importance in showing that the Moon is nearer to the Earth than Mars and is not interested in the date. We have calculated that the only occultation of Mars which was visible in Greece during Aristotle's lifetime, with the Moon at approximately first quarter, occurred in 357 BC (on 4 May). This must be the event to which Aristotle refers. It is interesting that, within the context of his account, Aristotle gives a very careful description (i.e. a rough estimate of the Moon's phase and the correct statement regarding the side of the Moon at which Mars disappeared and reappeared). However, the observation is of historical interest only, since no times are given.

1.2.5. Close Conjunction of a Planet and a Star

Ptolemy: Almagest, XI, 3. 'We have again chosen among the ancient observations one of the most certain, to determine the periodic motions of Jupiter. We learn that in the 45th year of the era of Dionysius, the tenth day of the month Parthenon, the planet Jupiter hid the south-east star of the Ass. Now this epoch coincides with the 83rd year since the death of Alexander, on the morning of the 17th to the 18th day of the Egyptian month Epiphi.'

The recorded date corresponds to BC 241 September 4. The star described is δ Cnc. The source of the observation is unknown, but the observation itself is reliable, as computation shows.

1.3. Mediaeval Europe (AD 500–1450)

Perhaps rather surprisingly, although there was little real interest in astronomy during the Middle Ages in Europe, numerous observations of certain kinds of phenomenon are recorded. Especially from the thirteenth century onwards, astronomical speculation flourished, notably among such writers as John of Sacrobosco, Ristoro d'Arezzo, St Thomas Aquinas and Nicole Oresme. However, this was largely cosmological in nature and the various writings contain very little in the way of actual observation. We must look elsewhere.

As far as astronomical records are concerned this was very much the era of the chronicler. The early period is rather sparsely covered by histories written by men such as Bede, Gregory of Tours, Cedrenus, Syncellus and Theophanes. Later, a few monastic annals give additional coverage. However, after about AD 1000 numerous monasteries, as well as some towns, began to keep chronicles. These annals are by far the most prolific sources of European astronomical records before the Renaissance. To give an illustrative example, Newton (1972b), who has made a detailed investigation of solar eclipse observations, published some 600 reports of these phenomena (not all independent) obtained from about 300 separate chronicles between AD 500 and 1300. Thus the main reason why we find so many observations from Europe at this period is because of the large number of individual sources. Most of the records come from England, France, Germany and Italy, where the concentration of monasteries was highest.

In general, the various chronicles are essentially concerned with local affairs (though this does not apply so much to the British annals as it does to those of continental Europe). Occasional reference to major national and international events is also made. As might be expected, natural disasters, such as earthquakes, floods, droughts and severe winters, are often reported. However, only the most spectacular celestial phenomena are alluded to — particularly solar and lunar eclipses and comets, and, to a lesser extent, bright meteors, or meteor showers, and aurorae. In almost every case, the observer was interested in the phenomenon purely as a spectacle: records of less striking events — e.g. occultations of planets and bright stars by the Moon or planetary conjunctions — are very infrequent. For a special reason, sightings of new stars (novae and supernovae) or sunspots are extremely rare. It would appear that scant notice was taken of these phenomena because their existence conflicted with contemporary cosmology, which, following Aristotle, supposed that outside of the immediate vicinity of the Earth the heavens were perfect and changeless. Comets escaped this censorship because they were believed to be atmospheric in origin.

We are indeed fortunate that extensive compilations of mediaeval European chronicles exist. Some of the best known are: *Rerum Britannicarum Medii Aevi Scriptores*, *Recueil des Historiens des Gaules et de la France*, *Monumenta Germaniae Historica, Scriptores* and *Rerum Italicarum Scriptores.* All of the above are multi-volume works. In general, these contain very few chronicles after about AD 1250, possibly because by this period there were just too many to publish systematically. Nevertheless, it is possible to build up a fairly complete list of observations made before AD 1250 from readily accessible sources.

The following are examples of the kinds of observation found in the records from mediaeval Europe.

1.3.1. Solar Eclipses

> *Leonis Deaconis Historicae.* 'AD 968 . . . . There was a defection of the Sun at the winter solstice such as never happened before . . . . The defection was such a spectacle. It was the 22nd day of the month December at the fourth hour of the day. The sky was clear when darkness was spread over the Earth and all the brighter stars revealed themselves. Everyone could see the disc of the Sun without brightness, deprived of light, and some dull and feeble glow, like a narrow band shining round the extreme edge of the disc. Gradually the Sun going past the Moon (for this appeared covering it directly) sent out its original rays and light filled the Earth again . . . . At the same time I myself was also staying in Constantinople . . . .'

This is the only clear description of the corona before the sixteenth century (see below). It was only in the eighteenth century that the phenomenon began to be studied in detail. The date of the observation is given precisely (on this day there was a total solar eclipse visible in Europe and the Near East). As the author is careful to say where he was at the time, and the report seems to be that of an eyewitness, it seems certain that totality was observed in Constantinople. It is an unfortunate fact that very few eclipse reports trouble to mention where the observation was made; usually this has to be presumed. The time of the eclipse is only given to the nearest hour, but this is fairly typical.

Ristoro d'Arezzo: 'Della Composizione del Mondo.' 'And while we were in the city of Arezzo, where we were born, and in which we are writing this book, in our monastery, a building which is situated towards the end of the fifth district, and whose latitude from the equator is 42 and a quarter degrees, and whose westerly longitude (from Baghdad?) is 32 and a third, one Friday at the sixth hour of the day, when the Sun was 20 degrees in Gemini, and the weather was calm and clear, the sky began to turn yellow, and we saw the whole body of the Sun covered step by step, and obscured, and it became night; and we saw Mercury close to the Sun, and all the stars which were above the horizon; and all the animals and birds were terrified; and the wild beasts could easily be caught; and there were some people who caught birds and animals, because they were bewildered; and we saw the Sun entirely covered for the space of time in which a man could walk fully 250 paces; and the air and the ground began to become cold; and it began to be covered and uncovered from the west.'

Although the year is not given, the month and approximate day can be deduced from the right ascension of the Sun ('20 degrees in Gemini'). In the thirteenth century this corresponded to around 4 June. Large solar eclipses were rather rare around this time, and the date readily reduces to AD 1239 June 3, a Friday. Ristoro would then be a young man. We have selected this particular record because the description is unusually detailed, although as in the previous example, the only *astronomically accurate* statement is that the Sun was totally eclipsed.

1.3.2. Sunspots

Niconovsky. '(AM 6879). During this year there was a sign in the Sun. There were dark spots on the Sun, as if nails were driven into it, and the murkiness was so great that it was impossible to see anything for more than 7 ft . . . . Woods and forests were burning and the dry marshes began to burn . . .' (translated by Vyssotsky 1949).

This Russian chronicle records two instances of sunspots, one in the year AM 6873 (AD 1367) and the second, quoted above, in AM 6879 (AD 1371). On both occasions the chronicle reports devastating forest fires, and the consequent dimming of the Sun apparently rendered the sunspots visible.

Observations of sunspots made elsewhere in Europe are virtually unknown.

1.3.3. *Lunar Eclipse*

> *Annales Sancti Blasii.* '1128 .... In the same year the Moon was turned into blood on the fifth day before the Ides of November.'

The recorded date corresponds to AD 1128 November 9. On the previous evening there was a total lunar eclipse visible in Germany. 'The Moon turned into blood' is a common expression in mediaeval European chronicles to describe a total eclipse of the Moon. A possible origin of this expression is in *Joel* ii, 31 (repeated in *Acts* ii, 20). However, the Moon often glows a deep red colour during totality on account of sunlight dispersed in the Earth's atmosphere, and this description may come most readily to mind. The above observation, like all other mediaeval European sightings of lunar eclipses, is of no real astronomical value.

1.3.4. *Planetary Conjunction*

> *Gervasii Monachi Cantuarensis Opera Historica.* '1170 .... On the Ides of September, at midnight, two planets were seen to be in conjunction with one another, with the result that it seemed as if they were one and the same star; but they were soon separated.'

The date is equivalent to AD 1170 September 13. Reference to the planetary tables of Tuckerman (1964) shows that the two planets were Mars and Jupiter. During the night of 12–13 September the planets passed within a minute of arc of one another, so that the unaided eye probably would not be able to resolve them. Here we have an instance where the observer, although giving a very careful description, was mainly interested in the phenomenon as a spectacle. It would appear that either he could not identify the planets concerned or he considered such a detail as of little importance.

1.3.5. *Supernova*

> *Annales Sangallenses Maiores.* '1006. A new star of unusual size appeared, glittering in aspect and dazzling the eyes, causing alarm. In a wonderful manner this was sometimes contracted,

sometimes diffused, and moreover sometimes extinguished. It was seen likewise for three months in the inmost limits of the south, beyond all the constellations which are seen in the sky.'

This report from Switzerland is the only detailed European observation of the supernova of AD 1006, which is by far the brightest supernova on record. There are only two certain European observations of this star (the other is from Benevento in Italy), which is the only nova or supernova reported outside the Far East before the Renaissance. The above report suggests that the star was at a very low altitude, even when on the meridian.

1.3.6. Comet

Annales Sanctae Columbae Senonensis. '1066 .... In this year with the shining Sun in the first part of Taurus, on the eighth day before the Kalends of May (24 April) there appeared a comet in the last part of the same (constellation), which was scattering sulphureous fires towards the south. When it began it extended as far as Saturn, also situated in the last part of Gemini, and from its very rapid motion it was understood to have been Mercury. However, after 15 days when the Moon was gleaming and coming close to it, it was soon extinguished.'

This French account represents one of the most detailed European observations of Halley's comet at its AD 1066 apparition. However, the writer reveals his very poor astronomical knowledge (and that of his contemporaries). There was no planet in or near Gemini and Saturn was more than 100 degrees east of the Sun. The nature of the object identified as Saturn (presumably a fixed star) is uncertain. Evidently the one characteristic of Mercury which was known to the various observers was its rapid motion; they seem to have been totally ignorant of its appearance.

1.3.7. Meteor Shower

Annales Seligenstadenses. '1122. Innumerable stars seemed to fall and as if to rain down over the whole Earth, on the day before the Nones of April.'

The date of this spectacular meteor shower corresponds to AD 1122 April 4. Another similar shower is reported in the same chronicle in April of AD 1095. Possibly both represent appearances of the April Lyrids.

1.3.8. Meteorite

> *Annales Sancti Blasii.* '1143 . . . . A fiery stone like a mass of burning iron fell from the sky onto Mount Brisach in front of the doors of the church.'

1.3.9. Aurora

> *Canonici Wissegradensis Continuatio Cosmae.* 'AD 1128 . . . . In the same year on the fifth day before the Ides of November . . . lunar eclipse . . . . Ten days having passed, at night a red sign appeared in the sky towards the north.'

The anonymous canon of Vysehrad (Prague) who continued the chronicle of Cosmas had a remarkable interest in astronomy and records numerous celestial phenomena.

The date of the lunar eclipse corresponds to AD 1128 November 9 and on the previous evening there was a total eclipse of the Moon visible in Europe. The probable date of the aurora referred to is thus the night of 18–19 November.

1.4. Renaissance Europe (AD 1450–1609)

From about AD 1450, reasonably accurate timings (to the nearest minute or so) of celestial phenomena are recorded by a number of astronomers. Many monastic chronicles were still maintained, recording events in a style which had not changed in more than half a millennium, but the later observations from these sources are of little more than academic interest.

At the beginning of this period, astronomers such as Regiomontanus and Walther were carefully timing solar and lunar eclipses in order to check the accuracy of their tables. Later observers of note were Copernicus, Gemma, Clavius and Maestlin. Just before the beginning of the telescopic era, when our survey stops, Tycho Brahe, Fabricius and Kepler, in particular, were making positional observations to a fraction of a minute of arc.

Somewhat surprisingly, the timed observations are of little value at the present day. This is because, as the result of extrapolation on slightly later telescopic observations (which are much more accurate, largely on account of the vastly increased resolving power available), it is possible to compute the times of eclipses far more accurately than they were observed. Only two observations in this period appear to be of any use in studying the variability of the Earth's rotation period and motion of the Moon (see chapter 2). Quite by chance, the Jesuit astronomer Christopher Clavius saw two total solar eclipses within a period of seven years. In the first of these (AD 1560) he happened to be very close to the southern limit of totality, and in the second case (AD 1567) the belt of totality was only some 10 km wide.

Again, by far the most interesting and useful planetary observations from this period are not carefully timed, but are mere sightings of extremely close conjunctions of planets with one another and with stars. There are a number of such instances in which, to the keen-sighted observers, an occultation appeared to be taking place. Such observations could be of importance in studying the motions of the planets in the past.

By this time, some of the best scientific minds in Europe were breaking free of the Aristotelian concept of a changeless celestial vault, and in many respects the careful observations of the supernovae which appeared in AD 1572 and 1604 represent some of the greatest triumphs of the age. Unfortunately, at this period there seem to be no sightings of sunspots (these had to await the application of the telescope to the study of astronomy).

In quoting examples of observations made at this period, we have restricted our attention to the more useful solar eclipse and planetary observations, and also some of the reports of the supernovae of AD 1572 and 1604.

1.4.1. Solar Eclipses

> *Clavius: In Sphaeram Ioannis de Sacrobosco.* 'I will cite two remarkable eclipses of the Sun which happened in my own time and thus not long ago; one of which I observed in the year 1559 about midday at Coimbra in Lusitania (Portugal) in which the Moon was placed directly between my sight and the Sun, with the result that it covered the whole Sun for a considerable length

of time and there was darkness in some manner greater than that of night. Neither could one see very clearly where one placed his foot; stars appeared in the sky, and (miraculous to behold) the birds fell down from the sky to the ground in terror of such horrid darkness. The other I saw in Rome in the year 1567, also about midday, in which although the Moon was again placed between my sight and the Sun it did not obscure the whole Sun as previously, but (a thing which perhaps never happened at any other time) a certain narrow circle was left on the Sun, surrounding the whole of the Moon on all sides.'

On the first occasion Clavius was a student at the University of Coimbra. He has mistaken the year, for the only total solar eclipse visible as such in Portugal between AD 1540 and 1600 occurred in 1560 (21 August). This eclipse would reach its maximum phase in Portugal just before noon, in keeping with Clavius' estimate. In the intervening time between this and the second eclipse, he had moved to Rome and was now teaching mathematics at the *Collegio Romano.* He was incredibly fortunate to witness two central eclipses in his lifetime, especially within a period of only seven years. Clavius presumed that the eclipse of AD 1567 (exact date 9 April) was annular. However, unknown to him, in the Mediterranean region, where the eclipse was visible near noon, the apparent diameter of the Moon slightly exceeded that of the Sun so that the eclipse was total. Elsewhere on the Earth, near the sunrise and sunset positions, it was annular. In the vicinity of Rome, the limb of the Moon would be no more than 3 arcsec outside the solar disc so that the inner corona would be very intense. This must be what Clavius is referring to when he speaks of the 'certain narrow circle' as 'surrounding the Moon on all sides.' Because very minor variations in the speed of rotation of the Earth would make the eclipse partial at Rome, this untimed observation is of considerable value.

Andreas Fosse, Bishop of Bergen and a mathematician, gave an account of the solar eclipse of AD 1601 December 24 as related by fishermen on the neighbouring sea coast. Although of academic interest only, because of the extremely wide zone of annularity, the account is worth reproducing because it is the first definite description of an annular eclipse from Europe.

Longomontanus: Astronomica Danica. 'They who had seen the aforesaid eclipse with very great astonishment described it. But in the same form the Sun had encompassed with its own circumference the Moon in the middle in such a way that its own light having been diffused up to the edge on all sides it more or less shone out equally by $1\frac{1}{2}$ digits.'

The above statement is very vague and untechnical but it is fairly clear that the fishermen witnessed a central annular eclipse. Possibly the eclipse was particularly striking since it occurred not long before sunset.

1.4.2. Close Planetary Conjunctions

Kepler gives a brief account of two apparent occultations involving planets only. Both were observed by his tutor Michael Maestlin of Tübingen, who was gifted with exceptionally keen sight (Kepler tells us that Maestlin could see 14 stars in the Pleiades).

Kepler: Astronomiae Pars Optica.
(i) 'Michael Maestlin of Tübingen and I saw Jupiter totally eclipsed by Mars in the year 1591 on 9 January. The fiery red colour of Mars showed that Mars was inferior (i.e. nearer the Earth).'
(ii) 'With regard to Venus and Mars, an experiment concerns the same Maestlin in 1590. On 3 October at 5 am Mars was totally occulted by Venus, with the white colour of Venus indicating that Venus was lower'.

Kepler gives the above in reverse chronological order and we have retained this order for reasons of continuity. The event in AD 1591 may well have been a real occultation of Jupiter by Mars, but in 1590 Venus would be some five magnitudes brighter than Mars and a close approach would probably suffice. Both observations would seem worth combining with the oriental observations of close planetary conjunctions (see below), since they could provide a valuable check on the planetary positions at various epochs in the historical past.

It is interesting to compare Maestlin's observations with the planetary positions interpolated from Tuckerman's tables. Shortly before dawn on 9 January in 1591 the calculated separation of Mars and Jupiter would be about 0·01 deg. Again around 5 am local time on 3 October in 1590 the calculated separation of Venus and Mars would be 0·01 deg. Both were certainly extremely close approaches.

1.4.3. *Supernovae*

It is an incredible coincidence that two bright galactic supernovae should appear only 32 yr apart (in AD 1572 and 1604) at a time when observers of the calibre of Tycho Brahe and Kepler were flourishing. It is most unfortunate that these supernovae appeared so soon before the telescopic era, for no galactic supernova has since been seen. However, present-day astronomers have reason to be grateful to Tycho and Kepler for the detailed observations of position and brightness that they made.

The supernova of AD 1572 appeared in the circumpolar constellation of Cassiopeia and readily attracted attention, for it was about as bright as Venus. Tycho Brahe collected virtually all of the important European observations of the star, including his own, in his *Astronomiae Instaurate Progymnasmata*. He measured the angular separation of the new star from Polaris and nine of the principal stars of Cassiopeia to the nearest ½ arcmin. His instrument appears to have had a small systematic error of about 3 arcmin, as discussed in chapter 3. The Englishman, Thomas Digges, was the only other person to measure the position of the star with any accuracy and, although he made fewer measurements than his more famous contemporary, his results would appear to be free from significant systematic error.

Probably the most fascinating observations of the supernova relate to its varying brightness. Most are due to Tycho; no other astronomer seems to have troubled to estimate the brightness after maximum. We have expressed the individual observations in the form of a table (table 1.1). Baade (1945) reduced the various estimates to magnitudes, and from these results it is possible to draw a detailed light curve.

The supernova of AD 1604, which appeared in the southern constellation of Ophiuchus, does not seem to have been much brighter than Jupiter; Kepler did for this star much

Table 1.1. European estimates of the brightness of the supernova of AD 1572.

Date	Brightness estimate
AD 1572 Nov 2	Not seen (probably would have been noticed if reasonably bright)
Nov 6	First detected (by Maurolyco and Schüler)
mid-November	About as bright as Venus
December	About as bright as Jupiter
AD 1573 January	A little fainter than Jupiter; much brighter than first-magnitude stars
Feb–Mar	Equal to brighter stars of first magnitude
Apr–May	Equal to stars of second magnitude
Jul–Aug	Third magnitude; similar to brighter stars of Cassiopeia
Oct–Nov	Equal to stars of fourth magnitude; in November very similar to κ Cas
Dec–Jan	Scarcely exceeded stars of fifth magnitude
AD 1574 February	Reached sixth and faintest magnitude
March	So faint that no longer visible

what his predecessor as Imperial Mathematician at Prague — Tycho Brahe — had done for the supernova of AD 1572. Both Kepler and Fabricius measured independently the position of the new star in relation to surrounding bright stars. Like Tycho, they aimed at an accuracy of better than 1 arcmin, but this accuracy they *achieved.*

This particular supernova is of special interest since it was first sighted about 20 days before maximum. Despite its inferior situation it was readily discovered since it appeared very close (about 3 or 4 deg) to the three planets Mars, Jupiter and Saturn, the first two of which were in mutual conjunction. Possibly because of the interest aroused by the previous supernova only 32 yr before, the new star began to be observed diligently all over Europe. As a result, the pre-maximum light curve is better established than for most extragalactic supernovae observed in modern times. David Fabricius, who is particularly noted for his discovery of the variable star Mira Ceti in 1596, and several other reliable observers, had looked at Mars and Jupiter on the evening of 8 October in 1604. Nothing unusual was noticed, which suggests that the new star was probably fainter than about magnitude +3. On the following evening it was discovered independently by two Italians — Altobelli and a certain physician whose name is unknown. By this time the star was about as bright as Mars. Over the next few days its brilliance increased rapidly.

After maximum only Kepler, of the European astronomers, seems to have observed the declining brightness of the star systematically. Kepler made a detailed study of the new star in his *De Stella Nova in Pede Serpentarii.* From the various brightness estimates, Baade (1943) was able to produce an accurate light curve. Recently Clark and Stephenson (1977) have improved this light curve by incorporating Korean observations.

Table 1.2. European estimates of the brightness of the supernova of AD 1604.

Date	Brightness estimate
AD 1604 Oct 8	Not seen
Oct 9	As bright as Mars
Oct 10	Somewhat brighter than Mars
Oct 11	Still brighter than on Oct 10
Oct 12	Almost as bright as Jupiter
Oct 15	A little brighter than Jupiter
Oct 20	Much brighter than Jupiter (about twice as bright)
AD 1605 Jan 3	Brighter than α Sco, much fainter than α Boo
Jan 13	Brighter than α Boo and Saturn
Jan 21	About as bright as α Sco, a little brighter than Saturn
End of Jan	As bright as α Vir
Mar 20	Not much brighter than ζ and η Oph
Mar 27	Not much brighter than ζ and η Oph
Mar 28	Not much brighter than η Oph
Apr 12	As bright as η Oph
Apr 21	As bright as η Oph
Aug 12–14	As bright as ξ Oph
Aug 29	About as bright as ξ Oph
Sep 13	Fainter than ξ Oph
Oct 8	Difficult to see; fainter or equal to ξ Oph; last sighting on this day

The European brightness estimates are summarised in table 1.2. A few clearly unreliable estimates are omitted for the sake of clarity.

1.5. The Far East

Oriental history, as distinct from legend, begins with China some time around 1500 BC. The earliest written records — divination texts from the Shang Dynasty — were discovered accidentally less than a century ago near An-yang in north-east China. Turtle shells and animal bones bearing a very primitive form of Chinese script were unearthed in large numbers. These were originally sold to apothecaries as 'dragon bones' to be ground up as remedies for various ailments. Fortunately, their historical significance was soon realised.

Before the Shang period, nothing definite is known of Chinese history; archaeological excavations have been unable to support the orthodox view of an advanced civilisation well before 1500 BC. We would know much more about the history of the country in the millennium succeeding the Shang had it not been for the widespread 'Burning of the Books' in 213 BC. This was instituted at the command of Ch'in Shih-huang, who unified China in 221 BC and became its first emperor. The object of the holocaust was to eradicate all memory of the former warring states which had vied with the now ruling state of Ch'in. As a result of this systematic burning and the destruction caused by the sacking of the Ch'in capital only seven years later, irreparable gaps have been left in the history of ancient China. The extensive archaeological excavations now in progress in the People's Republic have unearthed several early star maps and literary works, and it is hoped that further significant discoveries will be made in the near future.

On the credit side, the history of China after 200 BC is extremely detailed and extends almost uninterrupted into modern times. The principal sources of Chinese history (including astronomical records) are the dynastic histories — compendious works, often in many volumes, written up at some time after the fall of each dynasty. This writing of history became something of a ritual, and most such works were compiled by civil servants in the 'Bureau of Historiography' who had free access to the official records of the dynasty. Most dynastic histories contain an extensive treatise on astronomy, often in several chapters, and this contains numerous observations of all kinds of celestial phenomenon.

Korean history begins much later than that in China. Around 50 BC the various tribes of the peninsula were amalgamated, largely under the influence of Chinese immigrants, into three kingdoms (Silla, Paekche and Koguryŏ). The Chinese script was probably introduced at this time, but its use did not become widespread until the fourth century AD. Since its unification as a single kingdom (known as Koryŏ) in AD 936, Korea has seen only two dynasties. The Wang Dynasty lasted until AD 1392 and the Yi Dynasty for more than 500 yr after this, until the Japanese annexation of Korea in 1910. The official histories of the period of the three kingdoms and the Koryŏ Dynasty are both patterned on a typical Chinese history; in fact, the latter work contains an extensive astronomical treatise (in three chapters) which is indistinguishable from that in a Chinese history. Although the Korean alphabet (known as Hangul) was invented by King Sejŏng in the fifteenth century AD, it never found favour among scholars and all major historical works are written in classical Chinese.

Soon after AD 400, the Chinese system of writing was introduced to the island world, Japan, by way of Korea. However, it is not until the seventh century AD that reliable history commences. From earliest times down to the present day the nation has had only one imperial family. No history of Japan can be compared with the dynastic histories of China and Korea. The astronomical records are scattered in a variety of works such as privately compiled histories and diaries of courtiers. We are indeed fortunate that Kanda Shigeru (1935) took it upon himself to collect and classify the astronomical records of his country, thus making them generally available.

Joseph Needham, the distinguished historian of Chinese science, has remarked: 'Broadly speaking one can say that everything there (China) is either printed or lost.' This is a direct result of the very early invention of block printing (seventh century AD). This same remark is essentially true also for Korea and Japan. Additionally, classical Chinese was virtually the universal written language for all three countries, thus making the oriental astronomical records surprisingly accessible.

What of the records themselves? The An-yang oracle bones contain a few references to solar and lunar eclipses, but dating them is fraught with difficulties. There are also occasional mentions of stars and one *possible* allusion to a nova. Much later, the 'Spring and Autumn Annals', probably edited by Confucius (551–479 BC), contains more than 30 observations of solar eclipses (three of which are described as total) between 720 and 481 BC. The *exact* date is recorded in almost every case. This is certainly the earliest surviving *series* of eclipse observations (as distinct from isolated sightings) in any civilisation.

From the Han Dynasty (202 BC–AD 220) onwards, we find regular astronomical observations of all kinds recorded: solar eclipses in abundance, comets, novae, planetary conjunctions, occultations of planets and stars by the Moon, lunar eclipses, meteors, aurorae, sightings of Venus in daylight and sunspots. Later, in Korea and Japan essentially the same kind of observation was practised. Especially after about AD 1000, we frequently find the same phenomena reported in two or even three countries. Even as early as the Former Han Dynasty, the right ascension of the Sun (to the nearest degree) is specified for almost every solar eclipse, and the whole series of astronomical records from this period gives the impression of having been written up by someone who had access to the collection of data at the imperial astronomical office — the 'Astronomical Bureau.' This remark is also true of the astronomical treatises in the later dynastic histories.

Mention of the Astronomical Bureau brings us to the subject of the 'official' character of oriental astronomy. By the Han Dynasty, an astronomical office was established as a special sub-department in the 'Ministry of State Sacrifices.' Throughout subsequent Chinese (and later in Korean and Japanese) history until modern times, the Astronomical Bureau existed as an important government office.

The Astronomical Bureau had two main functions: the observation and interpretation of celestial portents and the maintenance of a reliable calendar. An imperial observatory was built at each of the various capitals of China, Korea and Japan. As the capital was moved (e.g. after the fall of a dynasty) so a new observatory was built. Obviously the preparation of the calendar was an important function, but very few of the observations which were made by the Bureau, apart from eclipses, had any calendar significance. Indeed, the principal objective of the Astronomical Bureau seems to have been the study of celestial omens. It is largely because of this that the vast number of observations of all kinds have come down to us.

Selected examples of the more important kinds of observation are as follows.

1.5.1. Solar Eclipses

The earliest reliable observations are reported in the *Ch'un-ch'iu* ('Spring and Autumn Annals'), the annals of the state of Lu. More than 30 solar eclipses are recorded in this period, mostly without any descriptive details, but allowing for minor errors of intercalation at this early period, the correct date is almost always given. Three eclipses are reported as total (in 709, 601 and 549 BC). The account of the earliest of these is:

> 'Third year of Duke Huan .... Autumn, seventh month, (day) *jen-ch'en*, the first day of the month. The Sun was eclipsed and it was total.'

The date reduces to BC 709 July 8, the day of occurrence of a total eclipse visible in China, if the cyclical day (*jen-ch'en*) is taken to be correct but an error of one month in the calendar month is assumed. In all probability, totality was observed at Chü-fu, the capital of Lu.

During the Han Dynasty in China, and again in the T'ang Dynasty, the right ascension of the Sun, expressed to the nearest degree, is almost always given for a solar eclipse. For some unknown reason this is extremely rare at other periods, and is virtually unknown in Japan and Korea. The following example relates to a solar eclipse which was almost total at

Lo-yang, the Later Han capital in AD 120. The source is the astronomical treatise of the *Hou-han-shu*, the official history of the Later Han Dynasty.

> 'Yüan-ch'ou reign period . . . . Sixth year, twelfth month, day *wu-wu*, the first day of the month. There was an eclipse of the Sun and it was almost complete. On the Earth it was like evening. It was 11 degrees in *Hsü-nu*. The woman ruler showed aversion from it. Two years and three months later, Teng, the Empress Dowager, died.'

The date reduces to AD 120 January 18, on which day there was a total solar eclipse visible in China. Although brief, the above record is a very careful description of an eclipse which was virtually total.

Observations of solar eclipses made in Japan are in general of little interest. From AD 1100 onwards, a number of eclipses are reported as total in Korea, but without any further information. In Japanese history the eclipses of AD 628, 975 and 1460 are all described as total, but only that of AD 975 is reported in any detail. A particularly vivid account is the following, taken from the *Nihon Kiryaku*, a privately compiled history of Japan written about AD 1028.

> '(Ten-en reign period, third year), seventh month, day *hsin-wei*, the first day of the month. The Sun was eclipsed. Some people say that it was entirely complete. During both the (double) hours *mao* (5–7 am) and *ch'en* (7–9 am) it was obscured. It was like the colour of ink and without brilliance. All the birds flew about in confusion and the various stars were all visible.'

The date is equivalent to AD 975 August 10, on which day there was a total solar eclipse visible in Japan. Various other accounts indicate a Kyōto source.

1.5.2. Sunspots

Sunspots are recorded from an early period in China, Japan and Korea. The earliest sighting, made in China, dates from 28 BC and after about AD 300 fairly regular observations are recorded. These data have not received the attention they deserve.

Most records mention only the occurrence of a 'black spot' or 'black vapour' within the Sun, but many make an allusion to size, usually by comparison with various fruits. In a number of cases the duration of visibility is reported, while frequently more than one spot was noticed. Two examples are:

> *Sung-shih* ('History of the Sung Dynasty'). 'Shun-hsi reign period, twelfth year, first month, (day) *kuei-szŭ*. Within the Sun there was a black spot as large as a date. From (day) *wu-hsü* until *kêng-hsü* within the Sun there was a black spot.'

> *Koryŏ-sa* ('History of the kingdom of Koryŏ'). 'Fifteenth year of King Myŏngjong, first month, (day) *chia-wu*. On the Sun there was a black spot as large as a pear.'

The date of the first Chinese observation is equivalent to AD 1185 February 10, while that of the Korean sighting corresponds to the very next day. Both descriptions must refer to the same spot, although the choice of fruits of radically different size is somewhat puzzling.

1.5.3. Close Planetary Conjunctions

The Far Eastern astronomers kept such a close watch on the motion of the planets and recorded so many conjunctions that it is only to be expected that a significant number of very close approaches of planets to one another and to stars should be noted.

One of the earliest observations involving Jupiter is recorded in the astronomical treatise of the *Hou-han-shu.*

> 'Sixteenth year (of the Yung-p'ing reign period), first month, (day) *ting-ch'ou.* Jupiter trespassed against *Fang Yu-ts'an;* the first star at the north was not visible. (On the day) *hsin-szŭ* it finally became visible.'

The record is very clear about specifying the star, for both *Fang Yu-ts'an* and the 'first star at the north' of *Fang* refer to β Sco. The date of disappearance of β Sco corresponds to AD 73 February 12 and the reappearance to 16 February, i.e. four days later. At the time Jupiter was moving very slowly (less than 0·03 deg per day) so that in four days it would move about 0·1 deg.

1.5.4. Comets

Numerous cometary observations are recorded in the Far East, but the most valuable are sightings of Halley's comet, the only known bright short-period comet. The following report from the astronomical treatise of the *Sung-shih* gives an extremely detailed account of the motion of Halley's comet at its AD 1066 apparition. The translation is taken direct from Ho Peng Yoke's (1962) catalogue.

> 'On a *chi-wei* day in the third month of the third year of the Chih-p'ing reign period a (*hui*) comet appeared at the *Ying-shih* (thirteenth lunar mansion). In the morning it was seen at the east measuring about 7 ft, pointing south-west towards the *Wei* (twelfth lunar mansion) and reaching *Fen-mu.* It gradually moved faster towards the east and became concealed when it approached the Sun. Until the evening on a *hsin-szŭ* day it appeared at the north-west but without its rays. The comet moved further eastward. Then there was a white vapour about 3 ft in width penetrating the *Tzŭ-wei* (enclosure) and the Pole Star, joining the *Fang* (fourth lunar mansion) and with both its head and its tail getting below the horizon. The comet moves further eastward, passed *Wên-ch'ang* and *Pei-tou* and penetrated the *Wei* (sixth lunar mansion). On a *jen-wu* day the comet retained its rays and measured over 10 ft (1 *chang*) in length and 3 ft in breadth. It was pointing north-east and then it passed *Wu-ch'ê.* The white vapour became branched, stretching horizontally across the heavens, and penetrated *Pei-ho, Wu-chu-hou, Hsien-yüan* and *Wu-ti-tso* and *Nei-wu-chu-hou* within the *T'ai-wei* (enclosure).
>
> It reached the *Chio,* the *K'ang,* the *Ti* and the *Fang* (first, second, third and fourth lunar mansions). On a *kuei-wei* day the comet measured 15 ft. It had a broom-like vapour and resembled a ten-peck measure. From the *Ying-shih* (thirteenth lunar mansion) it moved to the *Chang* (26th lunar mansion) passing altogether 14 lunar mansions. The comet and the vapour went out of sight after a total of 67 days.'

Few observations are quite so detailed as this one, but nevertheless, particularly at the hands of Kiang (1971), the oriental records of Halley's comet have helped considerably the study of the comet's motion in the past 2000 years.

1.5.5. Supernovae

Numerous 'guest stars' are recorded in Far Eastern literature, and the most probable interpretation of these is nova or supernova sightings. Only for the objects of long duration (say six months or more) is there any real justification for assuming a supernova nature and studies of such objects, outlined in chapter 3, have led to a deeper understanding of the supernova process. The following examples of records of supernovae are taken from Japanese and Korean history.

> *Meigetsuki* (diary of the Japanese poet–courtier Fujiwara Sadaie). 'First year of the Yōwa reign period, sixth month, 25th day, *kêng-wu*. A guest star appeared at the north near *Wang-liang* and guarding *Ch'uan-shê*.'

Fujiwara, who lived between AD 1180 and 1235, had a special interest in guest stars, and in AD 1230 he compiled a list of earlier sightings of such phenomena. This list contains no fewer than three likely supernovae — AD 1006, 1054 and 1181. The above date corresponds to AD 1181 August 7. Although there is no mention of the brightness of the new star, the positional information is extremely valuable in fixing its location.

> *Sŏnjo Sillok* (annals of the reign of King Sŏnjo of Korea). '(37th year of Sŏnjo, intercalary ninth month), (day) *kuei-wei*. In the first watch of the night a guest star was seen above the stars of *T'ien-chiang*. It was 11 degrees in *Wei* lunar mansion and distant 109 degrees from the pole. Its form was as large as Venus and its ray emanations were very resplendent. Its colour was orange and it was scintillating.'

The above account is one of many in the *Sŏnjo Sillok* which describes the supernova of AD 1604. On almost every day for several months we find a similar description, except that changes in the brightness of the supernova are noted. The date of the above entry is equivalent to AD 1604 October 28. Although the Korean measurements of position are far less accurate than those made in Europe, the brightness estimates form a valuable supplement to the European estimates, and allow a detailed light curve to be traced.

We shall conclude this section with a translation of a single year's observations recorded in a selected work (the *Koryŏsa*). This gives some indication of the kinds of phenomenon noted and the detail with which the observations were recorded. The year chosen (more or less at random) is the third year of King Chonjong (AD 1367–8).

'Second month, (day) *i-ch'ou*, the Moon entered *Nan-tou*. (Day) *kuei-yu*, five comets appeared, each 5 or 6 ft in length. Third month, (day) *wu-tzŭ*, the Moon was eclipsed. Fourth month, (day) *ting-szŭ*, a meteor appeared at *Ti* and entered *T'ai-wei*. (Day) *jen-shen*, a large meteor appeared at *Chüeh* and entered *Nan-tou;* it was red in colour. Seventh month, (day) *i-szŭ*, the Moon penetrated the stars of *Hsin*. (Day) *kêng-shen*, Mercury, Mars and Venus met at *Chang*. Seventh month, (day) *ting-mao*, there were three meteors as large as cups; one appeared at *Chuan-shê* and entered *Wu-ch'ê*, one appeared at *T'ien-ch'uan* and entered *Kou-ch'en* — both were red in colour: one appeared at *Pa-ku* — it was white in colour. Tenth month, (day) *ping-tzŭ*, a meteor as large as the half-Moon appeared at *I* and entered *Tou* and *Kuei* — it was red in colour. (Day) *wu-yin*, Mars trespassed against *Chin-hsien*.'

Some yearly entries are much longer than this, others much shorter or even non-existent. However, it is clear that interest was expressed in almost every type of celestial phenomenon.

1.6. Babylon

Babylonian astronomical records come from two distinct periods — a short interval covering the reign of a single king during the first half of the second millennium BC, and an incomplete span of about 600 years between 650 and 50 BC.

The first group is essentially only a single document and exists only in late copies. This document gives details of the first and last visibility of the planet Venus during the 21 yr reign of King Ammisaduqa of the First Dynasty of Babylon. The observations are of particular importance since, in principle, they should make it possible to date accurately the reign of this king (which was some time between 2000 and 1500 BC). This would considerably improve the accuracy of early Babylonian chronology, but as yet a unique date has not been established.

The second group is of considerable interest since the observations are so varied. Most of the observational texts in this later group are in the form of astronomical diaries. Although some texts are devoted to sightings of the Moon or one of the five bright planets, the information contained appears to be derived from the diaries themselves, i.e. these are the primary sources of information. The earliest known diary dates from about 650 BC, but the keeping of regular diaries probably began about a century earlier. Thus, in the second century AD, Ptolemy — in his *Almagest* — quotes a number of lunar eclipse observations made in Babylon from 720 BC onwards, while one existing tablet devoted to lunar eclipses cites an observation made in 731 BC.

The observations themselves are written in a very abbreviated form of cuneiform script on clay tablets. These were baked after completion. Such a material is far more durable than paper, although rather brittle, and most of the existing texts are in a very fragmentary condition. When the tablets were first discovered about a century ago they were collected in a very haphazard fashion. Almost the entire British Museum collection, by far the largest in the world, was purchased from antique dealers in Baghdad. Their exact provenance was unknown for decades. It was left to Sachs (1948) from a study of the contents of the texts to prove that the place of origin was the city of Babylon. Before about 380 BC most of the material is missing, but after this date a remarkably large proportion has survived the ravages of time.

Much of what we know about the interpretation of the astronomical texts is the result of the labours of Jesuit F X Kugler at the beginning of the present century. More recently O Neugebauer and A J Sachs have developed Kugler's work considerably.

The arrangement of a typical astronomical diary is very regular; in fact, the Babylonian astronomers seem to have been far more systematic in their observing programme than their Chinese counterparts. However, the Chinese were much wider in their outlook than the Babylonians, recording all types of celestial event. The Babylonian preoccupation with lunar and planetary phenomena reveals itself in the very small number of cometary observations (some six are known) and the complete lack of sightings of new stars. In making their observations, the Babylonians seem to have been motivated chiefly by a desire to obtain better ephemerides of the Moon and planets, but the ultimate objective was probably astrological.

In an astronomical diary, detailed observations are given for each month. The entries for a particular month always begin with a statement regarding the number of days in the previous month (invariably 29 or 30), followed by (weather permitting) a measurement of the interval between sunset and moonset on the evening when the young lunar crescent was first visible. The sighting of the crescent determined the start of the month. When the sky was cloudy this result was calculated by some unknown method, usually with fair results.

In the succeeding entries a variety of lunar and planetary observations are reported: conjunctions of the Moon with planets or ziqpu ('normal') stars (31 reference stars spread out at irregular intervals along the ecliptic zone); conjunctions of the planets with these same stars; and the more general planetary phenomena, such as heliacal settings and risings and stationary points. Around the middle of the month, the four minimum intervals between the rising and setting of the Sun and Moon (i.e. moonset to sunrise, moonrise to sunset, sunrise to moonset, sunset to moonrise) are recorded. Any time between the thirteenth and fifteenth day of the lunar month the occasional eclipse of the Moon is reported, often in considerable detail, with times measured to the nearest *uš* (the interval for the heavens to turn through 1 deg, i.e. 4 min) and magnitudes estimated to the nearest digit (one-twelfth of the Moon's apparent diameter). Solar eclipses are reported on the 28th or 29th day of the month, and, again, the observations are frequently very detailed. Near the end of a month, a measurement of the interval between moonrise and sunrise on the last morning that the waning crescent was visible is given.

Time intervals were presumably measured with the aid of a clepsydra, but this appears to have been subject to large drifts. Stephenson (1974) analysed a large number (77) of measurements of the four-monthly full Moon intervals. These were recorded on a single tablet covering the period 323–319 BC. Steady drifts varying from 4 to 27% were detected. It would appear that the only useful *timed* observations at the present day are of eclipses occurring close to sunrise or sunset (the usual reference points), for here the effect of drift is minimal.

Fortunately, calendar conversion presents few problems in

the later period. Before the reign of Seleucus I (commencing in 312/311 BC), years were expressed in terms of the reign of each individual king. From 311 BC onwards, the Babylonians adopted the era of Seleucus, so that years were then dated continuously. In order to make the beginning of the year correspond with the seasons, seven intercalary months were inserted every 19 yr, always after the sixth or twelfth month. Schoch, in Langdon and Fotheringham (1928), made a detailed study of the first visibility of the crescent Moon. Tables which he constructed for the prediction of this phenomenon were incorporated by Parker and Dubberstein (1956) in their investigation of late Babylonian chronology. Their tables allow the ready conversion of Babylonian dates from 626 BC to AD 75 to the Julian calendar with an error of no more than one day.

The usual names for the Sun, Moon and five bright planets found on the late texts are as shown in table 1.3. According to A J Sachs (private communication), the term for a comet is zallummu. The nature of this object is identified by its long duration (many days) and situation (usually well away from the ecliptic zone).

The following examples of eclipse records give some indication of the quality of observation in late Babylonian times.

(i) 'Year 182 (Arsacid), i.e. year 246 (Seleucid), Arsaces, king of kings . . . . Month IX, 14 . . . . When the Moon rose, two digits on the south side (were eclipsed). In 9 *uš* after sunset more than one-third of the disc (was eclipsed). 8 *uš* duration of greatest phase, until it began to become bright. In 11 *uš* . . . to south-west it became bright. 22 (?) *uš* (total duration). During this eclipse the sky was overcast . . . . During this eclipse a halo . . . . Jupiter

Table 1.3. Names used on late texts.

Sun	*šamaš*
Moon	sin
Mercury	gu_4-ud
Venus	dele-bat
Mars	AN
Jupiter	mul-babbar
Saturn	genna

and Saturn stood there .... The other (planets did not stand there). 4½ cubits (i.e. 9 deg) behind β Gem .... At 6 *uš* before sunset' (translated by T J Huber (private communication)).

The above record, obviously somewhat fragmentary, is contained on British Museum tablet 32845 which is, in fact, devoted to the eclipse. The date reduces to BC 66 December 28, on which day there was a partial lunar eclipse visible at Babylon.

(ii) '(Second year of Philip, sixth month), 28th day. At 3 *uš* before sunset there was an eclipse (of the Sun) ... north-west side ... it set eclipsed (?)' (translated by T J Huber (private communication)).

The above is an extract from an astronomical diary (British Museum tablet 34093). The date reduces to BC 322 September 9, on which day there was a solar eclipse visible at Babylon. This observation is potentially one of the most accurate of all Babylonian timed contacts since the interval between the beginning of the eclipse and sunset is so small — only some 12 min. The word order an-ge_6 *šamaš* (rather than *šamaš* an-ge_6) is customary for a predicted eclipse. However, this must be a mistake, since it is out of the question to presume that the Babylonian astronomers could predict a solar eclipse with anything approaching this accuracy.

(iii) '(Year 175 (Seleucid), month XII_2). Daytime of the 29th, 24 *uš* after sunrise, a solar eclipse beginning on the south-west side .... Venus, Mercury and the 'Normal Stars' (i.e. the stars which were above the horizon) were visible; Jupiter and Mars, which were in their period of disappearance (i.e. between last and first visibility) were visible in that eclipse ... (the shadow) moved from south-west to north-east. (Time interval of) 35 *uš* for obscuration and clearing up (of the eclipse). In that eclipse, north wind which ...' (translated by A J Sachs (private communication)).

The above remarkably accurate description of a total eclipse of the Sun is contained in an astronomical diary (British Museum tablet 45745). The date reduces to BC 136 April 15, on which day there was a total eclipse of the Sun. There is no comparable account until the eighteenth century, and the whole statement is a testimony to the observational skill which the Babylonian astronomers had achieved.

1.7. The Arab World

Astronomical activity in the Arab World did not commence until about two centuries after the introduction of Islam. The peak period for astronomical observation seems to have been between this time and AD 1000, when there are numerous measurements of the times of solar and lunar eclipses, planetary conjunctions, equinoxes, etc. These were, in the main, modelled on the observations made by the Greeks and recorded in Ptolemy's *Almagest*, which circulated widely at this period, but in general they were much more accurate.

Although possibly superior observers in many respects to the Chinese, the Arabs did not have the intellectual freedom of their Far Eastern counterparts. Thus, for instance, the occasional sightings of sunspots were attributed to transits of Mercury and Venus, when we know that no such events could have taken place. Again, the bright supernova of AD 1054 which produced the Crab Nebula was passed over in silence, although the much more brilliant one of AD 1006 was recorded by several independent observers.

Like the Greeks, the Arabs seem to have been motivated chiefly by a desire to improve the accuracy of tables of the motion of the Sun, Moon and planets. Hence, most observations which they made were of this form. A selection is as follows.

1.7.1. Solar Eclipses

> *Ibn Hayyan: Al Muqtabis*. '299 AH .... In this same year on Wednesday, the last day of the month Sawal, a total eclipse of the Sun occurred. Darkness covered the Earth and the stars appeared. The greater part of the people believed that the Sun had set below the horizon and got up for the Prayer of Sunset (*Magrib*). The shadow dissipated in a normal time behind the distant horizon.'

The date of the above observation corresponds to AD 912 June 17. The last section of the *Al Muqtabis* in which this observation appears is essentially a chronicle of Cordoba, the capital of the Arab dominions in Spain at the time. Although not a scientific observation, the account is of importance since totality is clearly expressed.

Ibn Iunis. 'Eclipse of the Sun observed at Baghdad. The Sun rose eclipsed to the extent of a little more than one-quarter of its surface.... We observed the Sun (by reflection) in water in a sure and distinct manner. We found at the end when no part of the Sun was eclipsed any more, and when its disc appeared complete in the water, the altitude 12 degrees in the east, less one-third of the division of the instrument, which was subdivided into thirds of a degree, thus making $\frac{1}{9}$ degree.'

The date of the solar eclipse mentioned by Ibn Iunis, who wrote about a century later, is equivalent to AD 928 August 17. This, like many others made in Baghdad and Cairo around this time, is a very careful observation. The method of using reflection in water to reduce the brightness of the Sun is particularly intriguing. The altitude measurement has an intrinsic accuracy equivalent to about half a minute of time.

1.7.2. Lunar Eclipse

Ibn Iunis. 'Eclipse of the Moon observed at Baghdad.... The beginning was at $10^h\,14^m$ of the night of Thursday, the middle at $11^h\,21^m$, the end at 9 minutes of the daytime of Thursday, all in seasonal hours.... Height of Sirius at the beginning, 31 degrees in the east; revolution of the celestial sphere between sunset and the beginning of the eclipse, determined by the astrolabe, approximately 148 degrees.'

The above observation, which was made at nearly the same time as the solar eclipse (date AD 927 September 14) is one of the most careful of all Arabian observations of lunar eclipses. Three different estimates of the time of beginning are provided. The use of clock stars as an aid to timing a lunar eclipse was common at this period.

1.7.3. Occultation

Egyptian manuscript. 'The Moon was in conjunction with Mercury and eclipsed it at the town of Qus. (The planet) remained eclipsed for about half an hour. This was at the second hour of the night of Tuesday, Muharram 2, in the year 672 (Hijra).'

The date of this rather rare observation (Mercury is seldom visible) corresponds to AD 1273 July 18. The time of the observation is only approximate — certainly not accurate enough to be of use at the present time.

1.7.4. *Close Planetary Conjunction*

> *Ibn Iunis.* 'I saw . . . a perfect occultation of the Heart of the Lion by Venus, the morning of the sixth day of the week . . . an hour before sunrise. Venus was, on the morning of the fifth day of the week, more than a degree from the Heart of the Lion; the morning of the seventh day of the week she was more advanced by the same quantity; and the morning of the sixth day of the week the Heart of the Lion was not visible.'

The exact place of observation is not mentioned (this would only be important insofar as it would influence the standard time of the event). Otherwise the observation seems to be a very careful one. The date reduces to AD 885 September 9. Whether the occultation was in fact a 'perfect' one is difficult to say in view of the extreme brilliance of Venus at any time. However, it should provide a fix on the coordinates of Venus with an accuracy of the order of one arcmin.

1.7.5. *Comet*

The following describes the AD 1066 passage of Halley's comet: see also the description from China quoted earlier.

> *al-Kāmil fī al-tārīkh* (458 AH). 'In the first ten days of Jumādā al-Ūlā (early April AD 1066), a large star appeared with a long tail in the east. Its width was about three cubits and it stretched to the middle of the heavens. It lasted until the 27th of the month and then disappeared. Then there appeared at the end of the same month in the west a star whose light surrounded it like the light of the Moon. The people were alarmed and greatly disquieted. When night fell, the star appeared with tails stretching towards the south. It remained for ten days and then disappeared.'

1.7.6. *Supernova*

The following represents the most careful Arabian observation of the supernova of AD 1006. The observation gives the date of discovery (indirectly), an estimate of the brilliance of the object, and a measurement of the celestial longitude (although not the latitude). The date of discovery reduces to AD 1006 April 30.

Alī ibn Ridwān: Commentary on the *Tetrabiblos* of Ptolemy. 'I will now describe a spectacle which I saw at the beginning of my studies. This spectacle appeared in the zodiacal sign Scorpio, in opposition to the Sun. The Sun on that day was 15 degrees in Taurus and the spectacle in the 15th degree of Scorpio. This spectacle was a large circular body, $2\frac{1}{2}$ to 3 times as large as Venus. The sky was shining because of its light. The intensity of its light was a little more than a quarter of that of moonlight. It remained where it was and it moved daily with its zodiacal sign until the Sun was in sextile with it (i.e. 60 deg away) in Virgo, when it disappeared at once.

All I have mentioned is my own personal experience, and other scholars from my time have followed it and came to a similar conclusion. The positions of the planets at the beginning of its appearance were like this: the Sun and Moon met in the 15th degree of Taurus; Saturn was 12° 11′ in Leo; Jupiter was 11° 21′ in Cancer; Mars was 21° 19′ in Scorpio; Venus was 12° 28′ in Gemini; Mercury was 5° 11′ in Taurus; and the Moon's node was 23° 28′ in Sagittarius. The spectacle occurred in the 15th degree of Scorpio. The ascendant of the conjunction when the spectacle appeared over Fustat of Egypt was 4° 2′ in Leo. Also the tenth (house which included most of) Taurus began at 26° 27′ in Aries.

Because the zodiacal sign Scorpio is a bad omen for the Islamic religion, they bitterly fought each other in great wars and many of their great countries were destroyed. Also many incidents happened to the king of the two holy cities (Mecca and Medina). Drought, increase of prices and famine occurred, and countless thousands died by the sword as well as from famine and pestilence. At the time when the spectacle appeared calamity and destruction occurred which lasted for many years afterwards . . .' (translated by M Y Tamar-Agha (private communication)).

Having given examples of historical astronomical observations over a wide spectrum, we are now in a better position to discuss their application.

2. *Solar Eclipses*

2.1. Introduction

The analysis of ancient observations of solar eclipses is a classic problem in astronomy. As early as 1695, their application in the study of the motion of the Moon by Halley led to the discovery of the acceleration of the lunar motion. Much nearer the present time, similar investigations brought about the realisation that the length of the day is gradually increasing. The problem is still a very important one at the present time.

Before discussing in detail the application of historical eclipse observations, let us begin by examining the various types of pre-telescopic observation which are of value in making deductions about $\dot{e}$ (the acceleration of the Earth's spin expressed in seconds of time per century^2) and $\dot{n}$ (the acceleration of the lunar motion in arcsec per century^2). All observations involve the Sun and/or Moon; there would appear to be no useful planetary data. As might be expected, there are few straightforward measurements of position at a specified time. Most observations are of conjunctions: eclipses, occultations, etc. The observations which are of value fall into six principal categories:

(i) records of the occurrence of central (total and annular) solar eclipses;
(ii) records of the occurrence of large solar eclipses (usually of unspecified magnitude);
(iii) estimates of the magnitudes of partial solar and lunar eclipses;
(iv) measurements of the times of first or last contact for solar and lunar eclipses;
(v) measurements of the times of conjunctions of the Moon

with stars; and
(vi) spring and autumn equinox observations.

Observations in categories (i) and (ii) have been made throughout the world at all periods from at least as early as 700 BC. However, they suffer from certain major drawbacks — in many cases the place of observation and/or date is in doubt. Categories (iii)–(vi) are more in the nature of scientific observations, and there are few problems concerning date or station location. On the other hand, the time span is much more limited and the intrinsic accuracy is, in general, fairly low.

In the ancient world, estimates of the magnitudes of eclipses were chiefly the prerogative of the Babylonians (750–50 BC), but there are a very small number of Greek estimates (200 BC–AD 200). A small number of similar observations were made by the Chinese (around AD 600) and the Arabs (AD 800–1000). In all cases, magnitudes were essentially guessed to the nearest twelfth or fifteenth of the solar or lunar disc, and this information is far from accurate.

Useful estimates of the times of beginning or end of solar and lunar eclipses only come from Babylon and Greece in the ancient world (substantially the same date ranges as for magnitudes), and from the Arab lands in mediaeval times. The Babylonian observations have yet to be analysed, but their accuracy is fairly high — to the nearest 4 min of time. However, as will be shown below, even such *apparent* precision (clock errors may be large) is considerably exceeded by the observations of central solar eclipses. The few Greek measurements were only made to about the nearest third of an hour. However, the mediaeval Arabian observations are of similar accuracy to those from ancient Babylon.

A small number of Greek observations of conjunctions of the Moon with stars is recorded in Ptolemy's *Almagest.* As was customary at this period (200 BC–AD 100) times are expressed only to the nearest third of an hour. The Babylonians made many similar observations, but for some reason (in marked contrast to the eclipse measurements) these were even less precise.

Coming now to the spring and autumn equinoxes, the exact

moment of the equinox is difficult to determine with precision, and this was particularly true before the advent of the telescope. Practically all of the surviving ancient observations which are of any use were made by one man (Hipparchus of Rhodes) in a brief period between 162 and 128 BC. During this interval, Hipparchus made 20 measurements of the times of equinoxes. Unfortunately, his accuracy was low: he expressed all results to the nearest 6 h. Nevertheless, this type of data is of particular value since the Moon is not involved, and thus a direct determination of $\dot{e}$ can be achieved (see below). About a thousand years after Hipparchus, the Arabs made a substantial number of equinox measurements, but with scarcely higher precision.

Most of the observations in categories (iii)–(vi) have been analysed recently by Newton (1970, 1972a). He has also carried out a detailed study of untimed observations of large solar eclipses (Newton 1970, 1972b), but in this work he did not make any distinction between central and partial eclipses. Before coming to the main theme of this chapter — the analysis of the observations of central solar eclipses — let us first discuss the nature of solar eclipses.

2.2. Nature of Solar Eclipses

A solar eclipse takes place whenever the Moon comes directly between the Earth and the Sun, so that the lunar shadow falls upon the Earth. An eclipse seen from within the penumbral shadow is termed 'partial' (figure 2.1) and the exposed portion of the Sun appears in the shape of a crescent. Viewed from within the umbral cone the eclipse is total;

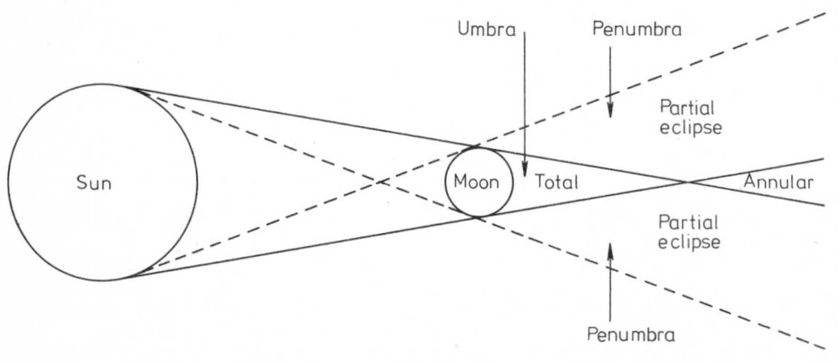

Figure 2.1. Shadow geometry for a solar eclipse.

here no portion of the solar photosphere is visible. An eclipse seen from directly beyond the vertex of the umbral cone is termed 'annular', because the Moon fails to cover a ring-shaped rim of sunlight.

By chance it so happens that at the Moon's mean distance from the Earth, the Sun and Moon appear to have roughly the same angular size. The Moon's orbit (around the Earth) is considerably more elliptical than that of the Earth (around the Sun): eccentricity 0·055 compared with 0·017. Thus, when the Moon is near perigee, central eclipses are invariably total, whereas near apogee they are annular. Because of the rather special geometrical circumstances prevailing, total and annular solar eclipses are of roughly equal frequency: each occurs about once every 1·4 yr on average on some part of the Earth's surface, but only about once every 300 yr at any one place.

To the unsuspecting observer, partial eclipses are seldom noticed, unless there is a thin layer of cloud or mist which shields the brilliant unobscured portion of the Sun, or, for example, someone happens to notice the eclipsed Sun reflected in a pool of water, etc. Muller (in Muller and Stephenson 1975), who has personally witnessed three total solar eclipses in recent years under good conditions, judges that in a clear sky even a 98% eclipse could easily go unnoticed; the thin crescent remaining would still dazzle the eye. This must be the main reason why, although annular eclipses are slightly more frequent than total ones, so very few are recorded. Not a single report from Europe before AD 1600 describes the annular phase unambiguously, yet more than 20 accounts clearly describe the total phase. On the other hand, given a thin cloud cover, even a small partial eclipse would be readily discernible; so, if there is no mention of magnitude in any particular instance, the actual phase could be large or small.

In many ways, an annular eclipse may be considered as nothing more than a special type of partial eclipse. In marked contrast, a total eclipse, in which no part of the solar photosphere is visible behind the Moon's limb, is possibly the most impressive natural spectacle. The fall of light intensity during the last few seconds is remarkably sudden (see figure 2.2) and

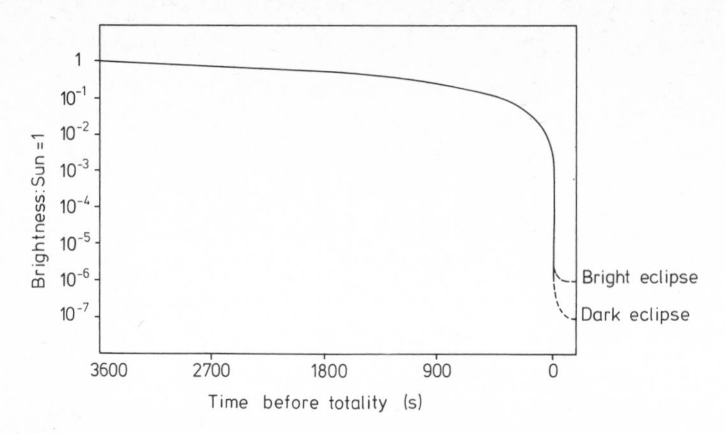

Figure 2.2. Schematic curve of change in light intensity at a total solar eclipse.

the shadow is seen racing in from the west at three times the speed of sound. During the total phase, the only light is provided by the dim solar corona (the extended outer atmosphere of the Sun) and scattered sunlight near the horizon from the region just outside the umbral shadow. Although the lunar limb is very irregular, any beads of sunlight shining down the clefts on the limb are so brilliant that even to the unaided eye it is clear that the eclipse is not quite complete. Thus in 1715, three observers in England who happened to be situated very close to the northern and southern limits of the path of totality reported as follows.

Darrington (north limit). 'The Sun . . . was reduced almost to a Point which both in Colour and Size resembled the Planet Mars; but whilst he (Mr Shelton) watched for the Total Eclipse, that Point grew bigger and the Darkness diminished; whence he argued the Limit to have been a very little more Southerly.'

Bocton (south limit). '. . . the Inhabitants . . . assured Mr Gray . . . that the Eclipse was not Total there, but, as one of them exprest it, before the Sun had quite lost his Light on the East Side he recovered it on the West: and there was a small Light left on the lower part of the Sun that appeared like a Starr.'

Cranbrook (south limit). '. . . William Tempest . . . observed there the Sun to be extinguished but for a moment and instantly to emerge again.'

The above quotations are from Halley's (1715) paper.

A highly original experiment, which might well be repeated with profit, was performed in New York City in 1925. The southern limit of totality for the eclipse of 24 January in that year passed through central Manhattan. In this experiment, 149 observations were made with the unaided eye by public employees who were stationed at one-block intervals along Riverside Drive and other streets roughly normal to the expected southern limit of totality. They took up their positions on the tops of buildings which were fairly low but which gave an unobstructed view of the oncoming shadow and the eclipsed Sun. Pairs of observers at each station were asked to note these two phenomena individually, and report their impressions. In particular, the observer was requested to decide whether, in his opinion, the eclipse was total or not total.

The results were impressive. All observations except one fell into two classes: those who reported totality and those who denied it. Only a single observer (very near the observed limit) expressed doubt. This remarkable consistency permitted the definition of the limit within some 200 m, and yet the observers were untrained personnel who had not witnessed a large eclipse before and who had no optical aid (for details see *TIES* 1925).

Most studies involving the application of solar eclipses in the investigation of the lunar motion and rotation of the Earth have concentrated largely, or entirely, on observations of total and annular eclipses. There are obviously sound reasons for the choice of the former. However, although the effects accompanying an annular eclipse are far less striking, there seems a reasonably good prospect that, if the eyes were shaded from the glare in some way (natural or otherwise), it would be possible to decide accurately whether the annulus was complete or not. Irradiation would have the effect of making even the narrowest ring of light seem wide. As the only two observations of annular eclipses are from China, and the Chinese maintained a careful watch for eclipses, these should be reliable enough. Paths of totality and annularity across the Earth's surface are usually fairly narrow (see figure 2.3). If we are confident that the total or annular phase is described in a particular record, then, no matter how $\dot{e}$ and

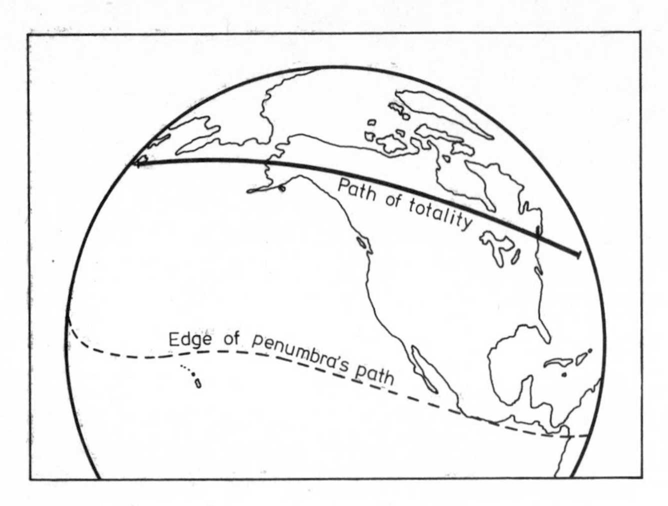

Figure 2.3. Path of the eclipse of 1963 July 20 across the Earth's surface.

ṅ are varied, the place of observation must remain within the narrow central zone.

Table 2.1 gives brief details for all those pre-telescopic observations of solar eclipses (of which we are aware) that satisfy the following criteria:

(i) the date is certain;
(ii) the place of observation is either certain, or at least identifiable with a good degree of confidence; and
(iii) the phase was either central, or, in a few special cases, nearly so, and in all instances the magnitude is clearly specified.

Under this last criterion, an observation of a total eclipse will be taken as one which in some way mentions the complete disappearance of the Sun. Allusion to darkness, or to the visibility of stars, is by no means a reliable guide to totality. In the Far East, a special term *chi* ('complete') was used to identify a total eclipse from very earliest times (at least from 709 BC). The character *chi* was originally a picture of a man kneeling by some food and turning his head away from it, indicating that he was replete. Although far from a scientific term, the meaning is perfectly clear when applied to an eclipse. Later (from 198 BC), the term was also used to denote an annular eclipse, but by this time it had become a technical expression indicating any central eclipse.

Table 2.1. Reliable pre-telescopic observations of central and near-central solar eclipses.

Date	Place	Description
BC 1375 May 3	Ugarit	Sun put to shame; went down in daytime
709 Jul 17	Chü-fu	Total (*chi*)
601 Sep 12	Ying (?)	Total (*chi*)
549 Jun 12	Chü-fu	Total (*chi*)
198 Aug 7	Ch'ang-an	Annular (*chi*)
181 Mar 4	Ch'ang-an	Total; dark in daytime
136 Apr 15	Babylon	Total; four planets, many stars seen
AD 65 Dec 16	Kuang-ling (?)	Total (*chi*)
120 Jan 18	Lo-yang	Almost complete; day became like evening
516 Apr 18	Nan-ching (?)	Annular (*chi*)
522 Jun 10	Nan-ching (?)	Total (*chi*)
840 May 5	Bergamo (?)	Sun hidden from world, then shone again
912 Jun 17	Cordoba (?)	Total; darkness covered the Earth
968 Dec 22	Constantinople	Sun deprived of light; clear account of corona
975 Aug 10	Kyōto	Total (*chi*); Sun the colour of ink
1124 Aug 11	Novgorod	Sun perished completely
1133 Aug 2	Salzburg	Sun suddenly disappeared; corona (?)
1133 Aug 2	Vysehrad	Partial; small crescent at south limb
1176 Apr 11	Antioch	Sun totally obscured
1178 Sep 13	Vigeois	Sun like a two- or three-day old Moon — just partial
1221 May 23	Kerulen River	Total (*chi*)
1239 Jun 3	Cerrato	Total
1239 Jun 3	Toledo	Sun lost all its strength
1239 Jun 3	Montpellier	Sun entirely covered by the Moon
1239 Jun 3	Arezzo	Whole body of Sun covered for several minutes
1239 Jun 3	Cesena	Sun covered with darkness; completely black
1239 Jun 3	Florence	Whole Sun obscured
1239 Jun 3	Siena	Sun completely obscured; no light

Table 2.1. *(continued)*

Date	Place	Description
AD 1239 Jun 3	Split	Whole Sun obscured
1241 Oct 6	Reichersberg	No part of the Sun could be seen
1241 Oct 6	Stade	Sun completely hidden from sight
1241 Oct 6	Cairo (?)	Complete darkness, other effects
1267 May 25	Constantinople	Total
1406 Jun 16	Braunschweig	Sun stopped shining
1415 Jun 7	Altaich	Sun entirely lost its light for several minutes
1415 Jun 7	Prague	Whole Sun eclipsed
1485 Mar 16	Melk	Complete eclipse
1560 Aug 21	Coimbra	Moon covered the whole Sun
1567 Apr 9	Rome	Narrow circle of light surrounded whole Moon

Certain observations describe a partial eclipse which was all but complete. We consider these just as reliable and useful, since they specifically deny that the central phase was witnessed. It might be pointed out here that, because the motion of the apsides of the lunar orbit (perigee and apogee) is so slow compared with the lunar motion — by a factor of several hundred — and the orbital eccentricity is fairly small, the lunar parallax can be computed very accurately, even in the distant past. There is thus no doubt about whether an eclipse was annular or total except in extremely marginal circumstances.

Referring to table 2.1, only in the very earliest observation is there any question regarding the date of occurrence. The year in which the phenomenon, which was almost certainly a total solar eclipse, took place is not recorded, but the date is given as 'the day of the new Moon in the month of Ḫiyar.' This corresponds to April–May. A computer investigation by Sawyer and Stephenson (1970) yielded BC 1375 May 3 as the only acceptable date. This was confirmed by Muller and Stephenson (1975).

In China, the presumed place of observation is the capital

(of the state of Lu in the earliest period, 720–480 BC, since the eclipses are reported in the annals of Lu, and later of the Empire). On any reasonable values of $\dot{e}$ and $\dot{n}$, totality is out of the question at Chü-fu in 601 BC and at Lo-yang in AD 65. On realising this, further historical research led to the choice of Ying — the capital of the then Presiding State in China — in 601 BC and Kuang-ling in AD 65. The eclipse announcement in AD 65 does not specifically mention Kuang-ling as the place of observation but associates it with the death of Ching, King of Kuang-ling (in Eastern China). Neither identification can be taken as certain, but each must be regarded as the most viable alternative to the original choice.

Only in a very few of the European observations is there any real doubt regarding the place of observation. Most reports are contained in monastic chronicles which are mainly concerned with very local affairs.

2.3. Analysis of Eclipse Observations

As stated earlier, it was an investigation by Halley which led to the discovery that the mean lunar motion is not uniform, but is accelerated. The mean longitude of the Moon is normally measured from the mean equinox of date. If there were no acceleration then the mean lunar longitude at any time T would be given by the expression

$$L = L_0 + aT.$$

It is customary to measure T in Julian centuries of 36 525 days from the epoch 1900.0 (1900 Jan 0, Greenwich mean noon). The more correct expression becomes

$$L = L_0 + aT + bT^2.$$

The quantity b has three principal components, one of which can only be determined by observation. It is this that concerns us here.

De Laplace (1788) made the first attempt at explaining the acceleration of the Moon. He showed that the diminution of the Earth's orbital eccentricity produced by planetary perturbations has the effect of accelerating the Moon. Adams (1853) showed that, although de Laplace's mechanism was correct, his result was seriously in error, and this was soon confirmed by Delaunay (1859). Brown (1919) derived the currently accepted figure of $6 \cdot 13$ arcsec century^{-2} for the coefficient of T^2. As this is derived from celestial mechanics, and the various parameters concerned are known accurately, the uncertainty should be small. Brown also deduced a small cubic term of coefficient $0''0068$.

Newcomb (1895) computed the annual precession in longitude of the equinox resulting from the action of the Sun, Moon and planets as $50''2371 + 0''0222T$ where T is in *tropical* centuries measured from the epoch 1850.0. His result for the acceleration of the equinox is thus $1''11T^2$. The motion of the equinox is in the direction of decreasing longitude, so that the coefficient of T^2 is thus $-1''11$.

Measured relative to the equinox we thus have

$$L = L_0 + aT + (7''14 + m)T^2 + 0''0068T^3$$

where m is either largely or wholly due to the effect of tides.

A variety of effects influence the rotation of the Earth. Tides raised by the Moon and Sun, both in the seas and the solid body of the Earth, cause dissipation of energy, and this tends to retard the Earth. This was first realised long ago by Kant (1754). The mechanism of lunar tidal friction is shown diagrammatically in figure 2.4. The Earth rotates faster on its axis than the Moon orbits the Earth, and the tidal bulges in the oceans thus have a phase *lag* (some 3 deg). This tends to

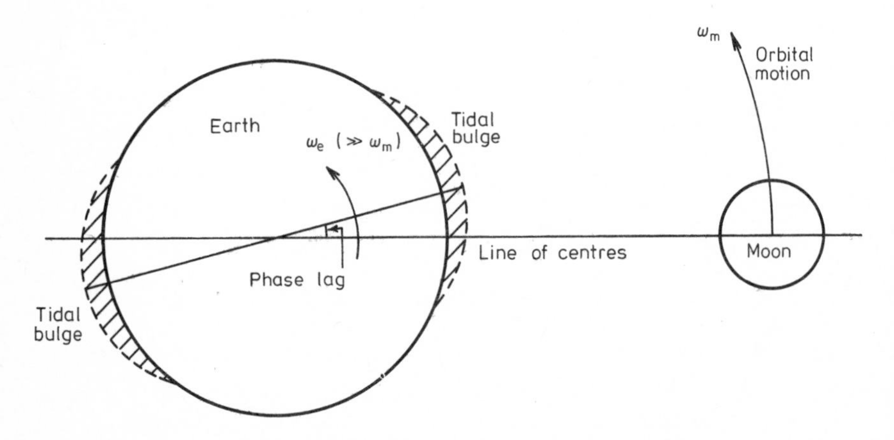

Figure 2.4. The lunar tidal phase lag.

retard the Earth's rotation, but accelerates the Moon in the direction of its motion. The Moon thus gains angular momentum at the expense of the Earth, but, because it recedes from the Earth, its angular velocity is actually decreased. The Sun has a similar effect on the Earth's rotation (of lesser magnitude), but the resultant acceleration of the Sun is completely negligible.

Numerous non-tidal causes accelerate the Earth's spin: changes in the moment of inertia of the Earth (e.g. in the polar caps, or in the interior of the Earth); electromagnetic coupling between the turbulent core and the mantle; solar-induced atmospheric pressure variations (semi-diurnal); etc. The resultant of the various tidal and non-tidal processes is a gradual slowing down of the Earth's rotation, but at an irregular rate.

The system of time known as Universal Time (UT) is based on the assumption that the length of the day is constant. Half a century ago, before the full realisation that the Earth's rotation is irregular, it was customary to speak of both the acceleration of the Sun and Moon (cf Fotheringham 1920, de Sitter 1927). The acceleration of the Sun was purely apparent, arising from the non-uniformity of the unit of time, but that of the Moon was partly apparent and partly real. Figures derived for the two accelerations were around +1·5 arcsec century^{-2} for the Sun and about +5 in the same units for the Moon (both expressed as coefficients of T^2). As the Moon moves on average some 13·4 times as rapidly as the Sun, the resultant real acceleration of the Moon on a uniform time scale, with no solar acceleration, was around −15.

Following work by Spencer-Jones (1932, 1939) which showed conclusively that the accelerations of the Sun and inner planets on UT are in the ratio of the mean motions, the first potentially uniform time scale, Ephemeris Time (ET), was adopted. On this scale the Sun has no acceleration and the Moon has an acceleration $\dot{n}$ (twice the coefficient of T^2). This is the time scale in which we shall be working. Defining T in terms of centuries of 36 525 *ephemeris* days (of constant length), we have for the Moon:

$$L = L_0 + aT + (7''14 + \dot{n}/2)T^2 + 0''0068\,T^3$$

where $\dot{n}$ is to be determined. The time difference

$$\Delta T = \mathrm{ET} - \mathrm{UT}$$

is a measure of the unreliability of the rotation of the Earth as a uniform source of time. It is basically the accumulated drift of the terrestrial clock. As will be shown below, it would appear, to a very good approximation, that

$$\Delta T = b + cT - \tfrac{1}{2}\dot{e}T^2$$

where b and c are constants.

As well as discussing the determination of $\dot{n}$ and $\dot{e}$ from pre-telescopic observations, one of the objectives of this chapter is to set some bounds on the departure of the ΔT curve from a simple parabola.

Figure 2.5 illustrates how inadequate is the assumption of a constant length of day in calculating the circumstances of an historical eclipse. Babylonian astronomers reported a total eclipse of the Sun on a date which corresponds to BC 136 April 15 (for full details see chapter 1). During totality, the four planets Mercury, Venus, Mars and Jupiter were seen as well as numerous stars, and the description is certainly the

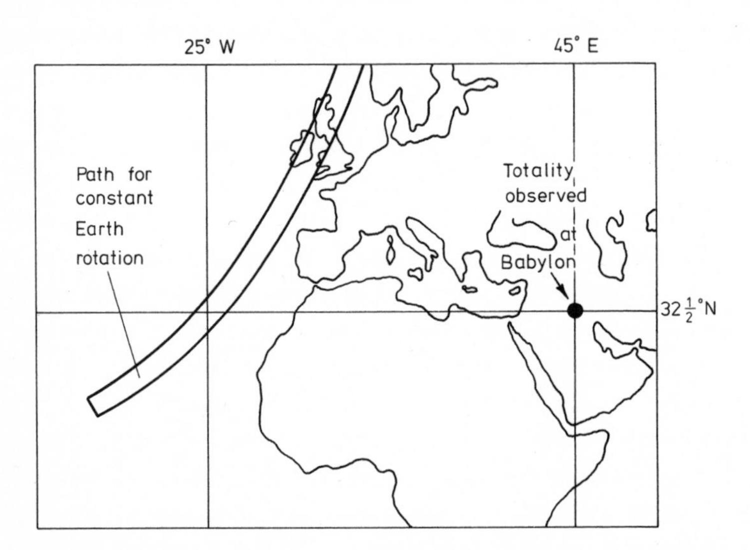

Figure 2.5. Earth's rotational displacement of a track of totality in the distant past as illustrated by the eclipse of BC 136 April 15.

most accurate account of a total eclipse at any time before the telescopic era. If we adopt a reasonable value for $\dot{n}$ (−35 arcsec century^{-2}), and calculate the path of totality on the assumption that $\dot{e} = 0$, the track comes out well to the west of Babylon. A rotation of the Earth of some 70 deg would be necessary to make the eclipse total at Babylon. The inference is that in the intervening time between 136 BC and the present day, the Earth has lost nearly five hours relative to an ideal clock.

In practice, there are two unknowns, $\dot{e}$ and $\dot{n}$. It requires several such observations to separate these parameters. Different authors have used varying techniques. Newton (1972b) preferred a least-squares solution capable of dealing with solar eclipses of any magnitude — not just total and annular. This is a very interesting method because there is no need to select only a very small number of observations of high reliability. Even very low-weighted observations can be analysed. However, a major objection to the use of a least-squares method was pointed out by Muller and Stephenson (1975). Several eclipses which crossed Europe during the mediaeval period were seen and recorded at a large number of independent centres. Thus the eclipse of AD 1039 August 22 was reported independently from some 20 sites (mainly monasteries) throughout central Europe. The path of annularity was extremely narrow and not a single station reported a central eclipse. At this comparatively recent epoch it is possible to predict the course of an eclipse to within a few tens of kilometres. This is not to say that the results obtained by different investigators for $\dot{e}$ and $\dot{n}$ are in good agreement, but the linear relationship between these quantities is determined far more precisely than the parameters themselves. Allowing only motion of the track of annularity in longitude, the least-squares solution is some 12·5 deg — and more than 1000 km — east of the true path (see figure 2.6).

The explanation is curious. In Europe at that time (as at any other time) the distribution of population centres — and of monasteries in particular — was non-uniform. Because what is now Germany had the highest concentration of monasteries, the least-squares path of annularity in 1039 is artificially displaced towards this area. Similar effects are

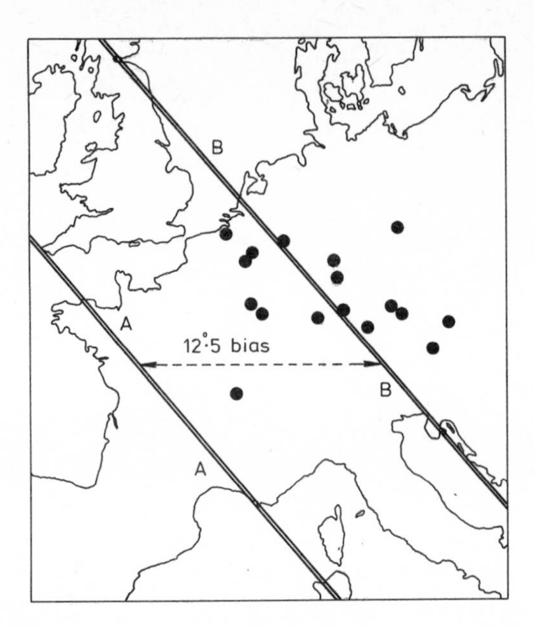

Figure 2.6. Population bias effects for the solar eclipse of AD 1039 August 22. Eclipse observations are indicated by the full circles. Path A is the actual path; path B is that implied by a least-squares fit to the observations.

observed at other eclipses — notably in AD 1093, 1187, 1191, 1239 and 1263. Population bias of such large magnitude can completely conceal any real trends in $\dot{n}$ and $\dot{e}$. Regretfully, because of this, and the great difficulty in assigning valid weights to individual observations (particularly when the reliability is obviously low), we feel that Newton's method of approach is unsuitable. In our view, the only satisfactory alternative is to use linear inequalities. This method makes use of only a small number of reliable observations of central, and near-central, solar eclipses.

The use of a method which rejects a large proportion of the available data might seem questionable, but we can find no application for the very large number of eclipse observations of unspecified magnitude. However, one definite advantage of selecting only a small body of data is that it is possible to enquire in detail into the historical circumstances surrounding each observation. This would not be practicable where a vast bulk of data was involved. The data given in table 2.1 are considered to be the most reliable pre-telescopic material for determining $\dot{e}$ and $\dot{n}$. In table 2.2, the geographical coordinates of the various places of observation are provided for reference.

Figure 2.7 shows the effect on the path of a central eclipse in the vicinity of a single place of observation produced by

Table 2.2. Coordinates of places of observation in table 2.1.

Place	Latitude	Longitude
Altaich	+48° 46′	−13° 02′
Antioch	+36° 12′	−36° 10′
Arezzo	+43° 28′	−11° 54′
Babylon	+32° 33′	−44° 25′
Bergamo	+45° 42′	−9° 40′
Braunschweig	+52° 15′	−10° 30′
Cairo	+30° 02′	−31° 15′
Cerrato	+41° 57′	+4° 31′
Cesena	+44° 08′	−12° 15′
Ch'ang-an	+34° 21′	−108° 53′
Chü-fu	+35° 32′	−117° 01′
Coimbra	+40° 13′	+8° 25′
Constantinople	+41° 01′	−28° 59′
Cordoba	+37° 53′	+4° 46′
Florence	+43° 46′	−11° 15′
Kerulen River	+48° 11′	−115° 54′
Kuang-ling	+32° 26′	−119° 27′
Kyōto	+35° 02′	−135° 45′
Lo-yang	+34° 47′	−112° 26′
Melk	+48° 14′	−15° 21′
Montpellier	+43° 37′	−3° 53′
Nan-ching	+32° 02′	−118° 47′
Novgorod	+58° 30′	−31° 20′
Prague	+50° 06′	−14° 25′
Reichersberg	+48° 20′	−13° 22′
Rome	+41° 54′	−12° 29′
Salzburg	+47° 48′	−13° 04′
Siena	+43° 19′	−11° 20′
Split	+43° 30′	−16° 27′
Stade	+53° 37′	−9° 29′
Toledo	+39° 52′	+4° 02′
Ugarit	+35° 37′	−35° 47′
Vigeois	+45° 23′	−1° 31′
Vysehrad	+50° 04′	−14° 25′
Ying	+30° 02′	−112° 15′

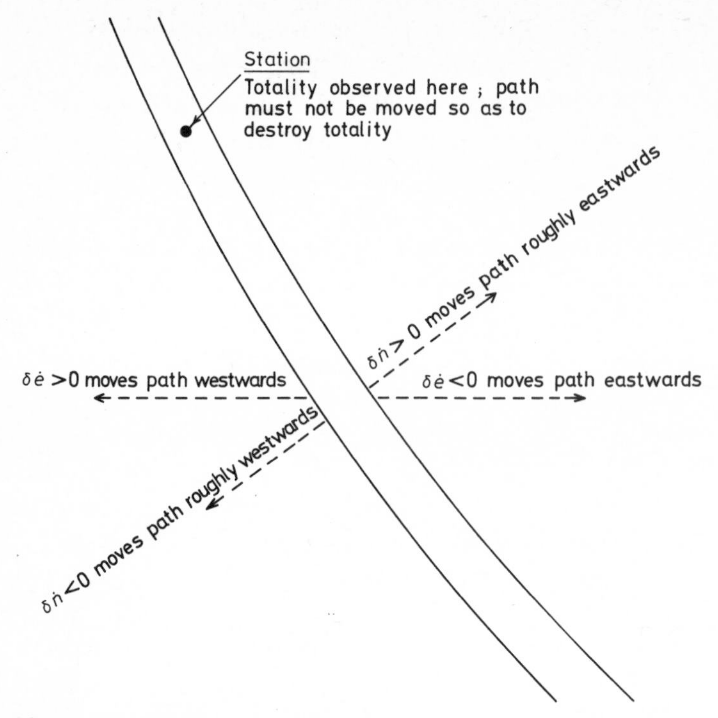

Figure 2.7. Effects on eclipse track by varying the accelerations of the Earth and the Moon.

small alterations in $\dot{e}$ and $\dot{n}$. A small change $\delta\dot{e} < 0$ moves the path precisely eastwards ($\delta\dot{e} > 0$ westwards), while a small change $\delta\dot{n} > 0$ moves the path only approximately eastwards, the precise direction depending on the circumstances prevailing. None of the pre-telescopic observations give any useful estimates of duration, or time of occurrence, of totality or annularity. If we are confident that an eclipse was central at a given place, then we can vary $\dot{n}$ and $\dot{e}$ independently in order to achieve this. Thus each observation provides a pair of linear equations. These correspond to the edges of the central zone just touching the place of observation. Where a central eclipse is expressly denied in the record, the same pair of equations is obtained, unless the observer has made it clear whether he was to the north or south of the central zone, in which case only one equation is appropriate. It will be seen that, if the effects produced by varying $\dot{e}$ and $\dot{n}$ were almost in the same ratio and same direction, there would be no way of separating these parameters. As it is, the sensitivity of the method is not as high as might be wished.

Figure 2.8 illustrates how various equations linking $\delta\dot{e}$ and $\delta\dot{n}$ can be combined in order to deduce the best solutions for $\dot{e}$ and $\dot{n}$ independently. Let $\dot{e}_0$ and $\dot{n}_0$ be preliminary solutions, and let $\delta\dot{e}_0$ and $\delta\dot{n}_0$ be departures from these The lines AA, A′A′ represent equations relating $\delta\dot{e}_0$ and $\delta\dot{n}_0$ corresponding to the edges of the central zone for some particular eclipse observation; similarly for BB, B′B′, etc. The best solution lies in the dotted area, that is it satisfies the various observations made on different occasions. Obviously, the more observations available the smaller the solution space becomes.

Unfortunately, some perfectly good observations (e.g. that represented by the bounds EE, E′E′) can fail to contribute anything to the final solution. This is because the solution space may well be small compared with the separation between the two bounds. Several precise observations in AD 1239 fall by the wayside for this reason. Figure 2.9 shows the approximate tracks of totality of the eclipse of 3 June across Southern France and Northern Italy. As the track of totality runs nearly along a line of latitude, even large changes in $\dot{e}$ and $\dot{n}$ do not destroy totality at Arezzo, Cesena, Firenze, Siena and Split. Essentially, all we can say is that there can be no doubting that the observations of totality made in these places are entirely confirmed. Similar remarks apply to the eclipses of 649 BC, AD 840, 975, 1176, 1406 and 1415.

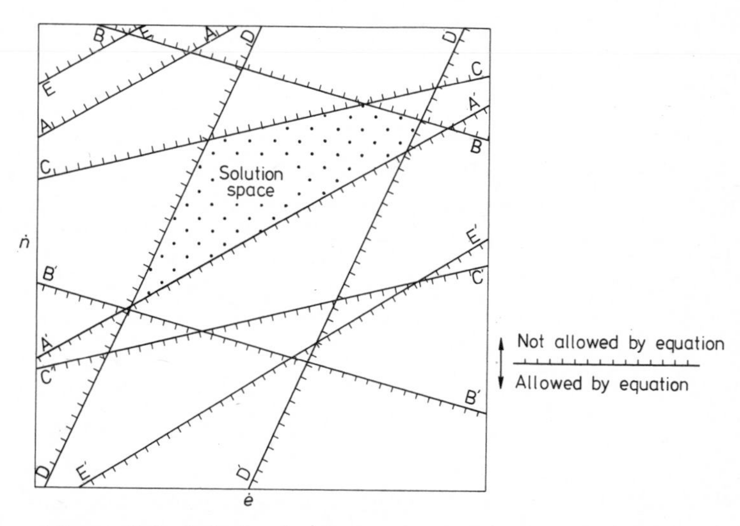

Figure 2.8. Solution by the method of linear inequalities.

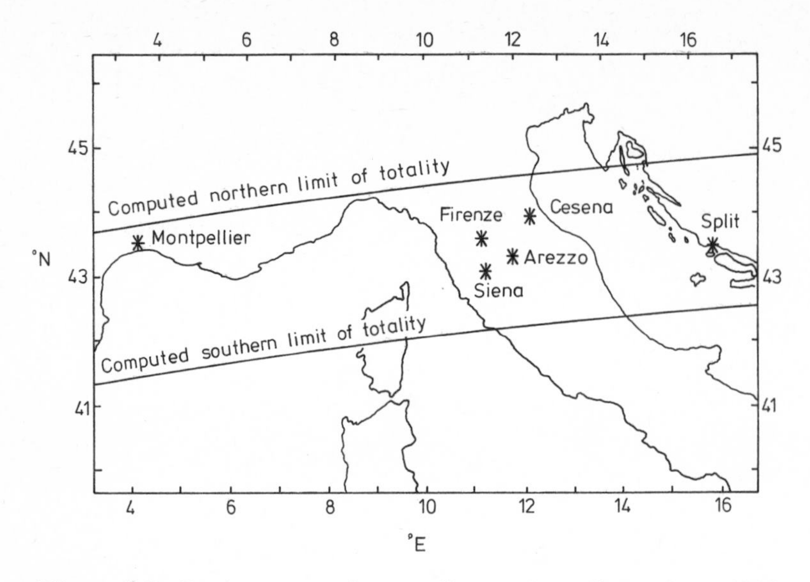

Figure 2.9. Stations reporting totality at the eclipse of AD 1239 June 3.

Once having obtained results for $\dot{n}$ and $\dot{e}$ by the above procedure, it is possible to study how ΔT (=ET−UT) has varied over the last 3000 years or so. Let us assume for simplicity that

$$\Delta T = -\tfrac{1}{2}\dot{e}T^2.$$

Figure 2.10 shows a hypothetical set of data covering the last three millennia. A result for $\dot{n}$ derived by the method just discussed is assumed. The abscissa represents the mean curve $\Delta T = -\frac{1}{2}\dot{e}T^2$ (T measured from 1900.0) and each vertical line gives the allowed range of ΔT ($\delta\,\Delta T$) at that epoch as determined from the relevant eclipse observation. In most cases, only a small portion of a vertical is shown: the eclipse tracks are normally so wide that *at most* only one limit is useful. Conversely, both limits are shown when a track is very narrow. What must be emphasised is that the limits are very hard, as in figure 2.9. Thus where a vertical does not intersect the abscissa there is definite evidence of fluctuation (i.e. a variation in ΔT from the mean parabola), unless there are grounds for questioning the reliability of the observation.

Before discussing the real ΔT curve, as deduced from the data in table 2.1, it is necessary to consider the nature of the

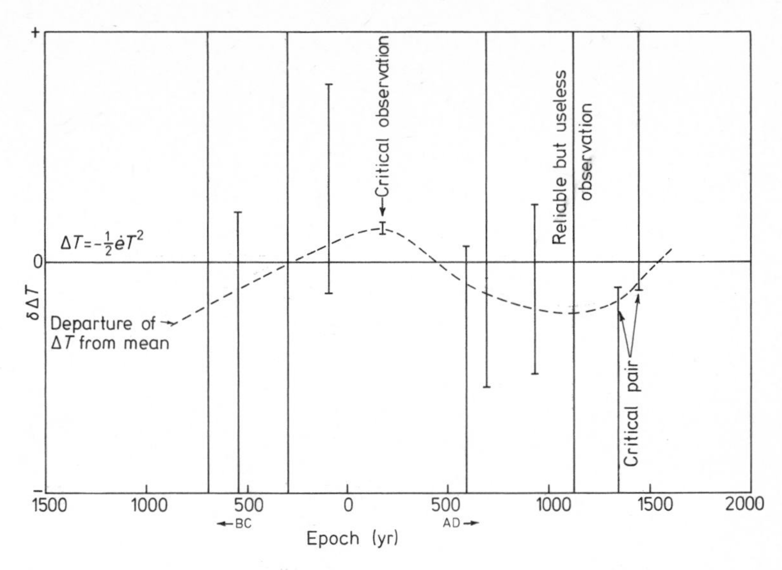

Figure 2.10. Schematic ΔT curve deduced from hypothetical data set.

ΔT expression in more detail. By definition, Newcomb's (1895) expression for the Sun's mean longitude is adopted as the foundation for ET. Spencer-Jones (1939) in his study of telescopic observations — occultations, meridian transits, and transits of Mercury across the solar disc — showed that, on his assumed value for the mean lunar acceleration on UT, the necessary correction to Newcomb's expression for the mean solar longitude (apart from fluctuations) was

$$+1''.00 + 2''.97\,T + 1''.23\,T^2.$$

As the Sun moves through 1 arcsec in an average time of 24·349 s, the corresponding result for ΔT was

$$+24^{s}.349 + 72^{s}.318\,T + 29^{s}.950\,T^2.$$

Here the result for $\frac{1}{2}\dot{e}$ is $-29^{s}.950$ century^{-2}, and thus $\dot{e} = -59^{s}.9$ in the same units. Both $b\,(=24^{s}.349)$ and $c\,(=72^{s}.318$ century$^{-1})$ — the constant term and coefficient of T — are substantial.

Recent investigations of historical astronomical observations have tended to use ET. However, when deriving a result for $\dot{e}$ (or its equivalent) it has been assumed that b and c are constant at the above values. Muller, in Muller and Stephenson (1975), showed that this rather arbitrary assumption was by no means justified. He developed an iterative method which enabled all three parameters b, c and $\dot{e}$ to be determined from the data. The final solutions were then

$$\dot{n} = -37{\cdot}5 \pm 5 \text{ arcsec century}^{-2}$$

$$\dot{e} = -91{\cdot}6 \pm 10 \text{ s century}^{-2}$$

$$b = +66{\cdot}0 \text{ s}$$

$$c = +120{\cdot}38 \text{ s century}^{-1}.$$

Obviously, neglect of considering possible alterations to b and c materially affects the final result for $\dot{e}$.

Figure 2.11 is based on figure 11 of Muller and Stephenson (1975). Here the value of $\dot{n}$ is taken as $-37{\cdot}5$ arcsec century^{-2}. The abscissa represents the mean ΔT curve

$$\Delta T_{\mathrm{m}} = 66^{\mathrm{s}}0 + 120^{\mathrm{s}}38T + 45^{\mathrm{s}}78T^2.$$

It can be seen that fluctuations from the mean are small and on the scale of the modern fluctuations observed accurately over the last 300 years or so. Muller in Muller and Stephenson (1975) attempted to fit the fluctuations in ΔT by a sinusoidal curve of amplitude 80 s and period roughly 1200 yr. This is purely speculative, but gives a useful indication of the magnitude and time scale of the allowed fluctuations. A number of reliable observations are not included

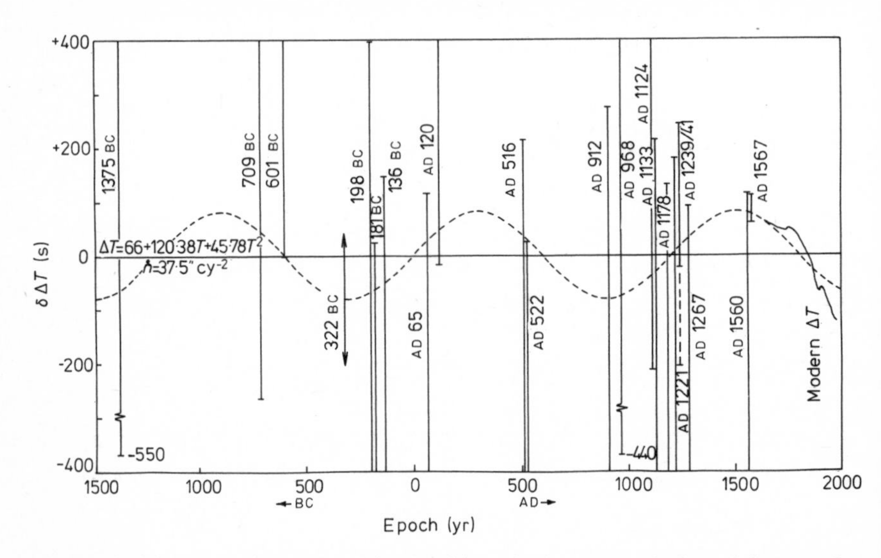

Figure 2.11. Variation of ΔT in the historical past as deduced from solar eclipses.

(e.g. several in AD 1239) simply because any value of ΔT within the bounds of the diagram would be acceptable.

The essential conclusion from figure 2.11 is that, aside from a steady increase in the length of the day over the last three millennia, short-term fluctuations in ΔT are of small amplitude (about 100 s or less). In recent literature it has become customary to use the quantity $\dot{\omega}/\omega$, where ω is the present rate of spin of the Earth, rather than $\dot{e}$. Conversion is readily made, since

$$\dot{\omega}/\omega = 3{\cdot}17 \times 10^{-10}\ \dot{e}.$$

The result for $\dot{e}$ of $-91{\cdot}56$ s century^{-2} thus corresponds to

$$\dot{\omega}/\omega = -29{\cdot}0 \times 10^{-9}\ \text{century}^{-1}.$$

A more meaningful way of expressing the result for $\dot{e}$ is in terms of the rate of increase in the length of the day. The derived value is $2{\cdot}50$ ms century^{-1}, equivalent to an increase in the length of the day by one second in 40 000 years at the present mean rate.

Hipparchus' equinox observations, which yield a direct result for ΔT, give at mean epoch 145 BC (Newton 1970)

$$\Delta T = 4{\cdot}2 \pm 1{\cdot}0\ \text{h}.$$

Using the value calculated from the derived expression

$$\Delta T = 66\overset{\text{s}}{.}0 + 120\overset{\text{s}}{.}38T + 45\overset{\text{s}}{.}78T^2$$

we obtain

$$\Delta T = 4{\cdot}65 \pm 1{\cdot}15\ \text{h}$$

in excellent agreement.

A matter of particular importance is the acceleration of the Moon. The result of $37{\cdot}5$ arcsec century^{-2} is equivalent to a tidal retreat of the Moon from the Earth of $5{\cdot}2$ cm yr^{-1}. It has long been known that there appears to be considerable discord between the results for $\dot{n}$ derived from ancient and modern (telescopic) observations.

Table 2.3 lists the results for $\dot{n}$ obtained by various authors. De Sitter (1927) essentially summarised all the early attempts based on pre-telescopic data.

Muller (1976) has shown that the apparent discrepancy may be attributed to the use of an incorrect motion of the

Table 2.3. Comparisons of various determinations of $\dot{n}$ from pre-telescopic and modern data.

	Authors	$\dot{n}$
Pre-telescopic data	de Sitter (1927)	$-37{\cdot}7 \pm 4{\cdot}3$
	Newton (1970)	$-41{\cdot}6 \pm 4{\cdot}3$
	Newton (1972b)	$-42{\cdot}3 \pm 6{\cdot}1$
	Muller and Stephenson (1975)	$-37{\cdot}5 \pm 5{\cdot}0$
Modern data	Spencer-Jones (1939)	$-22{\cdot}4 \pm 1{\cdot}1$
	Clemence (1948)	$-17{\cdot}9 \pm 4{\cdot}3$
	Oesterwinter and Cohen (1972)	$-38{\cdot}0 \pm 8{\cdot}0$
	Newton (1968)	$-20{\cdot}1 \pm 2{\cdot}6$
	Morrison and Ward (1975)	$-26{\cdot}0 \pm 3{\cdot}0$

lunar node in the investigation of historical observations. Martin and Van Flandern (1970) obtained a correction to Brown's (1919) motion of the node of $+4''{\cdot}31$ century^{-1}. If this value is adopted, Muller showed that the eclipse result obtained by Muller and Stephenson (1975) for $\dot{n}$ would reduce to $-30{\cdot}5$ arcsec century^{-2}. He concluded that it was 'premature to view the lunar longitude acceleration difference as a serious anomaly.' Certainly, there seems to be no geophysical basis for the assumption that $\dot{n}$ may have changed significantly during the historical past.

Extrapolation of the present tidal retreat of the Moon would bring the Moon very close to the Earth not much more than one aeon (10^9 yr) ago — considerably less than the ages of the oldest terrestrial and lunar rocks (up to 4·5 aeons). However, there is no guarantee that in the geological past tidal friction was as significant as it is now. Much depends on the distribution of the Earth's oceans, and particularly of the areas of shallow water. It seems quite possible that at the present time we may be experiencing a period of anomalously high tidal dissipation: after all, the time span of the historical data is only about one-millionth of the geological time scale.

A final point concerns the possible determination of a variation in G, the universal constant of gravitation. Van

Flandern (1975) made an interesting observational test based on modern and historical data. His analysis of lunar occultations over the period 1955 to 1974, utilising Atomic Time, gave a result for $\dot{n}$ of -65 ± 18 arcsec century^{-2}. He pointed out that this result differed significantly from other determinations which used ET. Taking what he considered to be the best determination of $\dot{n}$ on ET (-38 ± 4, the mean of several results), he concluded that the remaining acceleration (-27 ± 18) has as its most probable cause a decrease in G. His estimate for $\dot{G}$ was

$$\dot{G}/G = (-8 \pm 5) \times 10^{-11}\ \mathrm{yr}^{-1}.$$

A better figure for the ET result for $\dot{n}$ would be around -30 arcsec century^{-2}, but the uncertainty in the AT result is extremely large. Definite evidence for (or against) the non-constancy of G must await a better determination of $\dot{n}$ on Atomic Time.

3. *New Stars*

3.1. Introduction

In recent years there has been renewed interest in the pre-telescopic observations of 'New Stars' (galactic novae and supernovae). In particular, supernovae are of great astrophysical importance.

A supernova is the explosion signalling the dramatic self-destruction of a certain type of star which has reached the end of its normal evolutionary path. Such an event may result in the sudden implosion of the stellar core (on a time scale of seconds) to leave, in the case of total gravitational collapse, a black hole, or, when the collapse is halted at nuclear densities, a rapidly rotating neutron star which may be observable as a pulsar. The explosive ejection of the stellar envelope (representing a significant fraction of the stellar mass) produces filamentary structure flying off with velocities of thousands of kilometres per second. Finally, the outburst releases vast quantities of energy, of order 10^{51} erg (equivalent to about 10^{28} one-megaton hydrogen bomb explosions).

The last supernova detected in our galaxy was in AD 1604, although a large number of outbursts have been detected in nearby galaxies in the last few decades providing important new insights into the nature of supernovae. Since we lack telescopic observations of any galactic supernova, it is important that the historical records we do possess of such events should be analysed with great care.

Supernovae are of tremendous interest to astrophysicists, not only because they represent the most spectacular of stellar events, but also because the remnants and ejecta of such explosions are amongst the most unusual and exciting of astrophysical objects and phenomena. In addition to pulsars and black holes, supernova explosions are believed to be responsible for: the production of high-velocity 'runaway' stars hurtling through the Galaxy at speeds approaching a million miles per hour; the high-energy cosmic rays continually bombarding our planet; the heavy elements; spectacular expanding nebulosities which are amongst the most beautiful objects in the heavens; extended sources of radio emission; possibly gravitational radiation; and probably most of the galactic x-ray sources.

The blast wave expanding from the site of a supernova explosion delineates a region of synchrotron radiation (especially at radio wavelengths), and heated interstellar material swept up by the blast wave may be observed optically and in x-rays. The extended, long-lived (about 10^5 yr) radio source appears to be the most obvious remnant: in contrast to the easily obscured optical and x-ray emissions, radio waves may be detected from supernova remnants (SNR) on the remote side of the Galaxy. Less than a quarter of the 120 catalogued radio SNR have been detected optically, and a mere eight so far in x-rays.

The historical records of galactic supernovae give important information on the frequency of such outbursts, and on the development of their remnants. If a particular radio source can be identified with certainty as the remnant of a new star recorded in a particular year, then the time that the remnant has been expanding is known precisely. Several such identifications could then provide valuable observational evidence to test current theories on the nature of supernovae and the evolution of SNR.

By contrast with a supernova, which is a 'one-off' event heralding the final destruction of a star, a nova is believed to be merely a temporary departure of certain stars from the well established stellar evolutionary path, and may, in fact, be a repetitive phenomenon with an interval between outbursts of up to several thousand years. A prenova is believed to be a pair of stars close enough for mass transfer to occur between them (a *close binary* system). Any such system containing a white dwarf and a normal companion is a potential nova — transfer of hydrogen from the companion star to the white dwarf initiates a thermonuclear runaway process, with the explosive ejection of matter and the radiation of energy seen as the nova event.

A nova explosion resembles a supernova in many respects, but is on a much smaller scale. A typical nova lasts days, or at most weeks, rather than months, and the energy released is less than a ten-thousandth of that in a supernova explosion. In each nova explosion only a small fraction of the star's mass (possibly no more than one-hundredth of one per cent) is believed to be lost, and the ejecta dissipate in a few decades, leaving the binary showing little change from its original condition.

3.2. Historical Observations of New Stars

Most early observations of new stars were made in the Far East — China, Japan and Korea. In ancient and mediaeval Europe and the Arab Lands there seems to have been little interest in such phenomena, partly due to the widespread influence of the Aristotelian doctrine of a perfect changeless celestial sphere, and partly the result of sheer inability to recognise a new star-like object. In contrast, comets, which (again following Aristotle) were regarded as atmospheric phenomena, are recorded in abundance. It must be conceded, however, that a comet having a tail several tens of degrees in length can be a very spectacular object.

It would appear that the only useful observations of new stars, other than comets, made in the Western world before the Renaissance relate to the supernova of AD 1006. On the other hand, there is reason to believe that a substantial number of novae and supernovae are recorded in Oriental history.

Detailed catalogues of historically recorded new stars, giving durations, locations and descriptions have been compiled largely from original sources by Ho Peng Yoke (1962) and Hsi Tsê-Tsung and Po Shu-Jen (1965). Both catalogues contain Chinese, Japanese and Korean records. Additionally, the work of Hsi Tsê-Tsung and Po Shu-Jen contains a very small number of Western and Vietnamese observations. The catalogue of Ho Peng Yoke is particularly valuable since it lists comets as well as novae, etc, and makes no attempt to be selective. It is left to the reader to decide for himself the nature of a particular object.

Three terms were in common use in the Far East to describe new stars. These are:

K'o-hsing (Guest Stars), the normal expression to describe a fixed star-like object;
Po-hsing (Rayed Stars), which usually refers to tail-less comets; and
Hui-hsing (Broom Stars), the characteristic name for a tailed comet.

There was much confusion between the individual classes, since there are frequent accounts of moving *k'o-hsing*, while certain *hui-hsing* and *po-hsing* are recorded for many days without any reference to motion. Ho Peng Yoke (1962) has pointed out that the draft version of the Ming-shih, the official history of the Ming Dynasty in China, describes the supernova of AD 1572 as a *hui-hsing*, although the correct term *k'o-hsing* is applied in the history itself. Numerous other examples of such confusion could be given, and it is evident that caution must be exercised in the interpretation of the various descriptions.

Stephenson (1976) drew up a list of new stars considered to be possible novae or supernovae. Only those events for which there was no mention of motion, or clear reference to a tail, were included. The catalogue contains 75 stars, and is reproduced here as table 3.1. The columns given in order are:

(i) A reference number.
(ii) The Julian or Gregorian calendar date (the Gregorian calendar is used from AD 1582 September 15); frequently only the month is given.
(iii) An initial letter denoting the place of observation (A, Arab lands; C, China; E, Europe; J, Japan; K, Korea; V, Vietnam).
(iv) The recorded star type.
(v) The duration of visibility, where known.
(vi) A classification number from 1 to 5, depending upon the reliability (discussed below).
(vii) and (viii) The approximate right ascension (RA) and declination (Dec) (1950.0).

Table 3.1. Catalogue of pre-telescopic galactic novae and supernovae (from Stephenson 1976).

Ref. no.		Date	Place	Type	Duration	Class. no.	RA (1950.0)	Dec. (1950.0)	l	b	Remarks
01	BC	532 Spring	C	'star'	—	5	$20^h 50^m$	−10°	40°	−30°	
02		204 Aug–Sep	C	*po*	10 days	5	14 20	+20	20	+70	
03		134 Jun–Jul	C	*k'o*	—	4	16 00	−25	350	+20	
04		77 Oct–Nov	C	*k'o*	—	4	11 10	+75	130	+40	
05		76 May–Jun	C	*chu*	—	5	1 40	+25	135	−35	'Candle star'
06		48 May	C	*k'o*	—	4	18 40	−25	10	−10	
07		47 Jun–Jul	C	*k'o*	—	4	4 00	+65	140	+10	
08		5 Mar–Apr	C	*hui*	70 + days	2	20 20	−15	30	−25	
09	AD	61 Sep 27	C	*k'o*	70 days	2	14 10	+35	60	+70	Tail (?)
10		64 May 3	C	*k'o*	75 days	2	12 20	−5	290	+55	Tail (?)
11		70 Dec–Jan	C	*k'o*	48 days	1	9 40	+25	215	+45	
12		85 Jun 1	K	*k'o*	—	5	—	+65	—	—	Motion (?)
13		101 Dec 30	C	*k'o*	—	4	9 40	+25	215	+45	
14		107 Sep 13	C	*k'o*	—	4	6 30	+10	200	0	
15		125 Dec–Jan	C	*k'o*	—	4	17 10	+10	30	+25	
16		126 Mar 23	C	*k'o*	—	5	12 00	+10	270	+70	Motion (?)
17		185 Dec 7	C	*k'o*	20 months	1	14 20	−60	315	0	
18		222 Nov 4	C	*k'o*	—	4	12 30	0	290	+60	
19		247 Jan 16	C	*hui*	156 days	2	12 20	−20	295	+40	Tail (?)
20		290 Apr–May	C	*k'o*	—	4	—	+65	—	—	
21		304 Jun–Jul	C	*k'o*	—	4	4 20	+15	180	−25	
22		329 Aug–Sep	C	*po*	23 days	5	12 30	+55	130	+65	
23		369 Mar–Apr	C	*k'o*	5 months	1	—	+65	—	—	

25	393 Feb–Mar	C	*k'o*	8 months	1	17 10	+40	345	0	
26	396 Jul–Aug	C	'star'	50 + days	2	4 00	+20	175	−25	Reappeared later
27	402 Nov–Dec	C	*k'o*	2 months	2	11 10	+10	240	+60	Motion (?)
28	421 Jan–Feb	C	*k'o*	—	4	11 30	−15	275	+45	
29	437 Jan 26	C	'star'	—	5	6 40	+20	195	+5	Seen by day
30	483 Nov–Dec	C	*k'o*	—	5	5 30	0	205	−15	'Looked like *po*'
31	537 Jan–Feb	C	*k'o*	—	4	—	+65	—	—	
32	541 Feb–Mar	C	*k'o*	—	4	—	+65	—	—	
33	561 Sep 26	C	*k'o*	—	4	11 30	−15	275	+45	
34	641 Aug 6	C	*po*	25 days	5	12 20	+20	265	+80	
35	684 Dec–Jan	J	*po*	~2 weeks	5	3 40	+25	165	−25	
36	722 Aug 19	J	*k'o*	5 days	3	1 00	+60	125	0	
37	829 Nov	C	*k'o*	—	4	7 50	+15	205	+20	
38	837 Apr 29	C	*k'o*	22 days	3	7 00	+10	205	+5	
39	837 May 3	C	*k'o*	75 days	1	12 10	+5	280	+65	
40	837 Jun 26	C	*k'o*	—	5	18 00	−25	5	0	'Looked like *po*'
41	877 Feb 11	J	*k'o*	—	4	23 50	+20	105	−40	
42	891 May 12	J	*k'o*	—	4	16 50	−20	0	+15	
43	900 Feb–Mar	C	*k'o*	—	5	17 00	+10	30	+30	Tail (?)
44	911 May–Jun	C	*k'o*	—	4	17 10	+15	35	+30	
45	1006 Apr 3	A,C,E,J	*k'o*	Several years	1	15 10	−40	330	+15	
46	1011 Feb 8	C	*k'o*	—	4	19 20	−30	10	−20	
47	1035 Jan 15	C	'star'	—	5	1 20	+5	140	−55	
48	1054 Jul 4	C,J	*k'o*	2 years	1	5 40	+20	190	−5	
49	1065 Sep 11	C	*k'o*	—	4	9 20	−25	255	+20	Location doubtful
50	1069 Jul 12	C	*k'o*	11 days	3	18 10	−35	0	−10	
51	1070 Dec 25	C	*k'o*	—	4	2 40	+5	165	−50	
52	1073 Oct 9	K	*k'o*	—	4	0 10	+5	105	−55	

Table 3.1. (*continued*)

Ref. no.	Date	Place	Type	Duration	Class. no.	RA (1950.0)	Dec. (1950.0)	l	b	Remarks
53	1074 Aug 19	K	*k'o*	—	4	$0^h\ 10^m$	+5°	105°	−55°	
54	1138 Jun–Jul	C	*k'o*	—	4	1 50	+20	140	−40	
55	1139 Mar 23	C	*k'o*	—	4	14 10	−10	335	+50	
56	1163 Aug 10	K	*k'o*	—	4	17 30	−20	5	+5	Near Moon
57	1175 Aug 10	C	*po*	5 days	5	15 40	+50	80	+50	
58	1181 Aug 6	C,J	*k'o*	185 days	1	1 30	+65	130	0	
59	1203 Jul 28	C	*k'o*	9 days	3	17 10	−40	345	0	
60	1224 Jul 11	C	*k'o*	—	4	17 10	−40	345	0	
61	1240 Aug 17	C	*k'o*	—	4	17 10	−40	345	0	
62	1356 May 3	K	*k'o*	—	4	5 50	+30	180	0	Near Moon
63	1388 Mar 29	C	'star'	—	5	0 10	+20	110	−40	
64	1399 Jan 5	K	*k'o*	—	4	18 50	−20	15	−10	Near Moon
65	1404 Nov 14	C	'star'	—	5	19 50	+30	65	0	
66	1430 Sep 9	C	'star'	26 days	4	7 30	+5	215	+10	
67	1431 Jan 4	C	'star'	15 days	4	4 50	−10	210	−30	Reappeared later
68	1437 Mar 11	K	*k'o*	14 days	3	16 50	−40	345	0	
69	1460 Feb–Mar	V	'star'	—	5	11 30	−15	275	+45	
70	1572 Nov 8	C,E,K	*k'o*	16 months	1	0 20	+65	120	0	
71	1584 Jul 11	C	'star'	—	5	16 00	−25	350	+20	
72	1592 Nov 28	K	*k'o*	15 months	1	1 20	−10	150	−70	
73	1592 Nov 30	K	*k'o*	4 months	1	0 50	+60	125	0	
74	1592 Dec 4	K	*k'o*	3 months	1	0 00	+60	115	0	
75	1604 Oct 8	C,E,K	*k'o*	12 months	1	17 30	−20	5	+5	

(ix) and (x) The approximate galactic coordinates *l* and *b*.
(xi) Remarks.

The only new stars sighted in the West which are included in the catalogue are the well known supernovae of AD 1006, 1572 and 1604. There are no reliable European or Arabian observations of other new stars†. From the Far East the bulk of the new stars catalogued were originally classified as *k'o*. These include stars of both long and short duration, as well as those of indefinite duration for which there is only a mention of the first sighting. All isolated sightings of *po* are excluded. Use of this term implies that the star possessed a definite form, and when no duration is recorded the probability of it being a nova seems low. It is likely that a number of *k'o* of indefinite duration were comets, but here at least the 'correct' term for a nova is applied. Where a *po* is recorded with a definite duration, but without any hint of motion or tail, there is a reasonable possibility that the star was mis-identified. With two exceptions (in 5 BC and AD 247) all *hui* are excluded from the catalogue. These two stars remained visible for many days and there is no hint of any motion. All other such stars for which no motion or tail is mentioned were visible for about 25 days or less (usually much less), and accordingly seem scarcely worth considering as possible novae or supernovae.

Generally speaking, it is obvious that the reliability of the observations included in the catalogue varies enormously.

Referring to table 3.1, it is curious that no durations between about 25 and 50 days are reported. This provides a means of selecting the most promising supernova candidates. As shown by Clark and Stephenson (1976), any star of duration less than about 40 days can be virtually ruled out as a supernova. This is based partly on the evidence that there are only two possible references to the long-period variable star Mira Ceti in Far Eastern history (AD 1070 and 1592) before its discovery in Europe by David Fabricius in AD 1596. This star has a period of roughly 11 months. At maximum it reaches a magnitude of between +2 and +4 and at minimum

† Very recently, Brecher *et al* (1978) have discovered a Near Eastern sighting of the AD 1054 supernova.

descends to about +9. It is thus only visible to the unaided eye for about half of the time, and yet at bright maximum it is the brightest star in its neighbourhood. At a declination of −3°, it is very well placed for general observation. Taking conservative estimates of magnitude +3 as the minimum brightness for detection and +5·5 as the lower limit of unaided-eye visibility, judging from the type I and type II light curves of Barbon *et al* (1974), a supernova would take at least 40 days, and possibly much longer, to decline over this range of brightness.

From the above argument, we should expect to find any supernovae among the longer-duration stars (more than about 50 days), while stars of short duration (less than about 25 days) are not worth considering seriously as supernovae. This, of course, is not to imply that all stars of longer duration are supernovae: we might expect to find both slow novae and comets among this material. However, there can be little doubt that the short-duration objects are a mixture of novae and comets only.

Referring to the classification in column (v) of table 3.1, stars in classes 1 and 2 are all objects which were seen for more than about 50 days. Class 1 objects are all *k'o*, for which there is no hint of any tail or motion in the record. Stars in class 2 are either downgraded *k'o* (because of possible reference to a tail or motion) or stars of another type. Basically, class 3 objects are *k'o* of short, but definite,

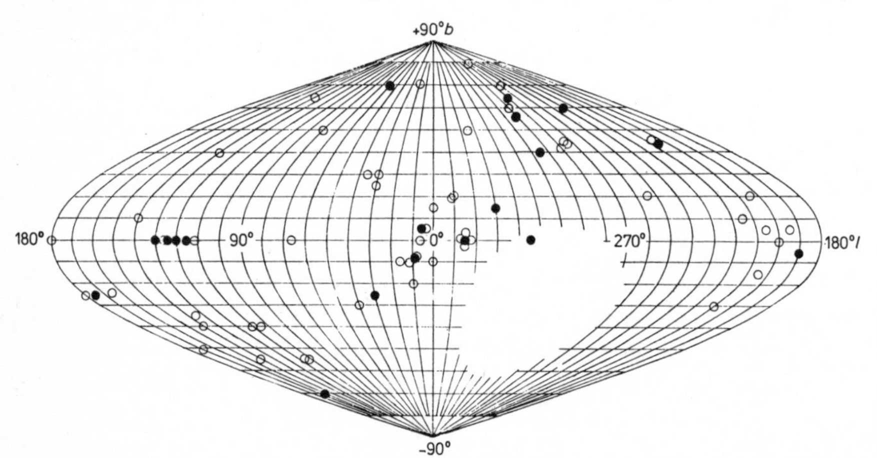

Figure 3.1. Galactic distribution of pre-telescopic novae and supernovae.

duration (up to a maximum of about 25 days). Class 4 stars are *k'o* of indefinite duration, or unspecified stars of known duration. Finally, objects in class 5 are unspecified stars of indefinite duration, or *po* of known duration. An object is downgraded by one class if there is any suggestion of a tail or motion.

It seems likely that most of the stars in class 3 were novae. However, the proportion of comets in classes 4 and 5 may be very significant. Figure 3.1 shows the positions of all of the stars in the catalogue in terms of galactic coordinates (l and b). The blank zone in the lower right of the figure represents the area currently not visible from 35°N latitude (the approximate location of the various Far Eastern capitals). Stars of medium to long duration (more than about 50 days) are represented by full circles, others by open circles.

The isotropic distribution of the short-duration objects is that which would be expected both for comets and for novae close enough to stand a reasonable chance of being discovered. (Stephenson (1976) has estimated an average distance for novae bright enough to be noticed of about 500 pc. From Payne-Gaposchkin (1957) the average distance from the galactic plane in the vicinity of the Sun is 275 pc. Thus a fairly isotropic distribution in galactic longitude and latitude is expected.)

It is obviously not possible to give a useful estimate of the proportion of novae in the catalogue. As stressed earlier, it could contain a few incorrectly classified comets when the historical records make no mention of motion or a tail. All that can be done is to select the most probable novae and supernovae, and leave the nature of the remaining objects in doubt.

Attention has sometimes been drawn to the possibility of two seemingly independent new stars representing separate outbursts of the same (recurrent) nova. In view of the poor positioning of such stars in general, and, with a few notable exceptions, the marginal evidence that any particular star was a nova, no suggested recurrent novae among the pre-telescopic objects can be taken seriously. None of the bright novae detected since the advent of the telescope have obvious historical counterparts.

The search for supernovae will be restricted to the 20 class 1 and 2 new stars in table 3.1.

Extragalactic surveys show that, in spiral galaxies, supernovae belong to a flat, rather than a spherical, population. Within our galaxy, SNR tend to lie close to the galactic plane. A source count from the Clark and Caswell (1976) catalogue of SNR gives the following distribution in galactic latitude: 75 sources for $|b| \leqslant 1°$; 34 sources for $1° < |b| \leqslant 5°$; 7 sources for $5° < |b| \leqslant 10°$; and only 4 sources for $|b| > 10°$. Although most galactic SNR surveys have been restricted to within a few degrees of the plane, the concentration of remnants towards the plane is not believed to be a consequence of this. Indeed, Henning and Wendker (1975) have concluded from a search of available high-latitude radio data that there are no additional high-latitude sources which could definitely be considered as SNR. The distribution of 'young' pulsars confirms that they were born close to the plane (Taylor and Manchester 1977). In the light of the above evidence, one must therefore conclude that, in the absence of a detectable remnant, any high-latitude new star was probably a nova. Thus we can rule out the following as likely supernova candidates: the stars of 5 BC (possibly a nova — Clark *et al* (1977b) have suggested that the record of this event may represent an independent sighting of the Star of Bethlehem), AD 61 and AD 64 (both possibly comets), AD 70 (possibly a nova), AD 247 (possibly a comet), AD 396 (probably a nova with secondary maximum), AD 402 (possibly a comet), AD 837 (very probably a nova, since its position remained fixed) and AD 1592A (perhaps Mira Ceti). Two maxima are seemingly recorded for the star AD 1592A, about a year apart. Mira has a period of roughly 11 months, but the recorded position of the star is more than 10 deg away. Additionally, the position of the star of AD 369 is recorded so poorly that we cannot even deduce an approximate galactic latitude. The probable declination of this star was around +65°, but the right ascension is in considerable doubt. There is no SNR north of +50° to which it could possibly correspond.

Both the other stars of AD 1592 appeared in Cassiopeia: 1592B in the neighbourhood of α Cas and 1592C close to β Cas. This area has been well surveyed for SNR, and none are known, apart from the intense young radio remnant Cas A. Brosche (1967) and Chu (1968) suggested independently that Cas A is the remnant of the star AD 1592C, but the preferred age of the remnant is much less than this (Gull 1973). There seems little doubt that both stars were slow novae (like that of AD 837).

We are thus left with the stars of AD 185, 386, 393, 1006, 1054, 1181, 1572 and 1604, and we will discuss these in reverse chronological order.

Kepler's supernova, which appeared in the constellation Ophiuchus in the year 1604, was carefully observed in Europe, China and Korea. The position of the star was measured so accurately by Kepler and Fabricius that Baade (1943) had no difficulty in identifying the optical remnant. Furthermore, the European and Korean astronomers, by comparing the brightness of the supernova with that of the neighbouring stars and planets, have made it possible to reproduce a remarkably precise light curve of the event. The radio remnant is the powerful source G4.5+6.8 (galactic longitude followed by galactic latitude), exhibiting the peripheral brightening characteristic of SNR (Gull 1975).

Tycho's supernova of 1572 was also observed both in Europe and in the Far East, but it is essentially only Tycho's observations that provide useful information. The position he determined for the object, although displaying an apparent small systematic error (Stephenson and Clark 1977), allows an identification with the radio source G120.1+1.4. The radio SNR again shows a symmetrical shell (Duin and Strom 1975), the position of the outer edge of the shell corresponding to the position of optical filaments. Tycho's SNR has also been observed in x-rays.

The supernova of 1181, unlike Kepler's supernova (peak apparent magnitude −2·5) and Tycho's (−4), does not seem to have been particularly brilliant — at its brightest it was about magnitude zero. It was recorded only in China and Japan, and its location with respect to three neighbouring asterisms in Cassiopeia was carefully described. Its position is in excellent agreement with the radio SNR G130.7+3.1.

The supernova of 1054 was again recorded only by the Chinese and the Japanese. From the combined reports we learn that the star was visible in the daytime for 23 days and that it appeared close to the third-magnitude star ζ Tau. The possibility that the supernova was associated with the Crab Nebula (G184.6−5.8) was first suggested by Lundmark (1921), although this identification has been criticised by Ho Peng Yoke *et al* (1970) who noted a directional inconsistency in one of the primary Far Eastern descriptions. This criticism can be discounted (Clark and Stephenson 1977); the expansion rate of the visible filaments of the Crab Nebula and the rate of increase in the period of its central pulsar indeed confirm that the Crab Nebula is one of the few known supernova remnants the age of which has been determined precisely.

The supernova that appeared in 1006 is the only one known to be recorded in both European and Arabian records before the Renaissance. It was also carefully observed by astronomers in China and Japan. Almost every source comments on the extreme brilliance of the star. A Chinese record states: 'It shone so brightly that objects could be seen clearly by its light.' One Arab record states: 'Its rays on the Earth were like the rays of the Moon.' Another comments: 'Its light illuminated the horizon and . . . its brightness was a little more than a quarter of the brightness of the Moon.' A Chinese record indicates that the star was observed for several years. From an analysis of the available historical records Stephenson *et al* (1977) confirmed that the remnant of the outburst is the radio source G327.6+14.5 (also detected at optical and x-ray wavelengths). At a peak magnitude of −9·5, it was the most brilliant supernova ever recorded, and must have been at a distance of only about 1 kpc.

There is no mention of the brightness of the AD 393 new star in the single record we have (from China), but, based on the fact that the star was visible for a period of eight months, we have inferred that it was probably brighter than magnitude −1 at maximum. It was described as being within the asterism *Wei*, which is the tail of the western constellation Scorpio — three SNR on the near side of the Galaxy fit this description. Thus, unfortunately, it is not possible to define uniquely an SNR in this instance. In the analysis to follow, we use the most likely candidate (G348.5+0.1) suggested by Clark and Stephenson (1977).

The short-duration new star of AD 386 shows a possible positional coincidence with the SNR G11.2–0.3. However, the evidence for it being a supernova remains, at best, circumstantial; the short duration and unknown brightness of the event leaves open a nova interpretation.

The last probable supernova on our list is the guest star of AD 185. From the solitary Chinese record of the new star, Clark and Stephenson (1975) have inferred that its likely remnant is G315.4−2.3 (previously suggested by Hill 1967), that the new star was observed for a total of 20 months after reaching a maximum apparent magnitude of −8, and that it must have occurred within 2 kpc of the Sun.

Of the pre-telescopic new stars recorded in history, there thus appear to be four certain supernovae, four probable, or possible, supernovae, and six likely novae. For every other star in the present catalogue we cannot be sure whether we are dealing with a nova or a comet. Further progress in the discovery of historically recorded temporary stars is likely to be very slow, since most of the well known sources have now been thoroughly searched.

The properties of the historical supernovae are summarised in table 3.2.

3.3. Application of Supernova Observations

We will now apply the data from the historical supernovae to a study of the evolution of SNR, and to estimate the frequency of galactic supernovae.

The picture of the evolution of SNRs which emerges from theoretical investigations suggests that, after a few hundred years of near-free expansion, the evolution resembles that of an adiabatic blast wave created by releasing energy at a point in a homogeneous gas. The diameter D (pc) of the shock wave preceding the expanding shell of swept-up interstellar material is then described by the Sedov (1959) similarity solution

$$D = 4{\cdot}3 \times 10^{-11} (E_0/n)^{1/5} t^{2/5} \qquad (3.1)$$

where t (yr) is the time elapsed since the explosion, E_0 (erg) is

Table 3.2. Properties of the historical supernovae.

Year	Constellation	Magnitude	Duration	Radio remnant	Other remnants	Records	Remarks
AD 185	Centaurus	−8	20 months	G315.4−2.3	Optical	Chinese	Probable SN
386	Sagittarius	?	3 months	G11.2−0.3?	None	Chinese	Possible SN
393	Scorpius	−1	8 months	G348.5+0.1?	None	Chinese	Possible SN
1006	Lupus	−9·5	Several years	G327.6+14.5	Optical and x-rays	Arabian Chinese Japanese European	Certain SN
1054	Taurus	−5	22 months	G184.6−5.8	Optical, x-rays and pulsar	Chinese Japanese	Certain SN
1181	Cassiopeia	0	6 months	G130.7+3.1	Optical	Chinese Japanese	Probable SN
1572	Cassiopeia	−4	18 months	G120.1+1.4	Optical and x-rays	Chinese Korean European	Certain SN
1604	Ophiuchus	−2·5	12 months	G4.5+6.8	Optical	Chinese Korean European	Certain SN

the energy released in the outburst, and n (cm^{-3}) is the number density of H atoms in the interstellar medium.

Statistical investigations of the radio remnants of supernovae by Milne (1970), Downes (1971), Ilovaisky and Lequeux (1972), and Clark and Caswell (1976) have confirmed that this functional dependence of D on t is observed in practice for the majority of SNR. Additional data are required to determine the constant of proportionality: one method is to estimate the rate of occurrence of supernovae using all those with known ages; another method uses the individual sizes and ages of historically recorded supernovae; and a rather different approach utilises the x-ray data for the older supernova remnants.

From considerations of all three approaches, Clark and Caswell (1976) concluded that the best estimate was $E_0/n \simeq 5\times10^{51}$ erg cm^3. Clark and Culhane (1976) found a slightly smaller value from a detailed consideration of the third method alone. However, both investigations are consistent with the relationship

$$D \simeq 0{\cdot}93t^{2/5}. \tag{3.2}$$

It is of interest to see how the six historical supernovae with certain, or probable, SNR identifications fit the evolutionary track implied by equation (3.1). The current linear diameters of Kepler's (AD 1604) and Tycho's (AD 1572) supernovae are 13·8 and 9·3 pc, respectively. Then, from equation (3.1), $E_0/n \simeq 21\times10^{51}$ erg cm^3 for Tycho's supernova and $3{\cdot}4\times10^{51}$ erg cm^3 for Kepler's supernova.

The suggested radio remnants for the supernovae of AD 1181 (3C58) and AD 1054 (the Crab Nebula) are atypical, in that both show central brightening and flatter-than-average non-thermal spectra. Because of other enigmatic differences, the discussion of the Crab Nebula is deferred until later. In the case of 3C58, despite the fact that it does not display the peripheral brightening characteristic of SNR, adoption of the outer boundary of radio emission as the shock-front diameter leads to $E_0/n \simeq 2{\cdot}8\times10^{51}$ erg cm^3, in surprisingly good agreement with most other remnants.

Interpretation of the historical data for the supernovae of AD 1006 and AD 185 requires that they be, respectively, at

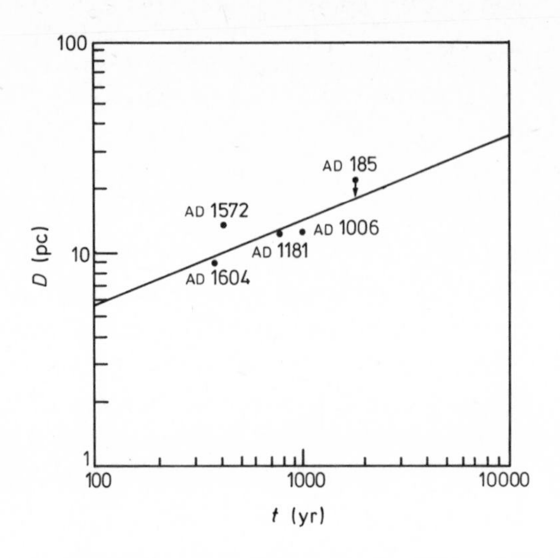

Figure 3.2. The D–t relationship.

approximately 1 kpc and up to 2 kpc from the Earth. The value for AD 1006 is then $E_0/n = 2{\cdot}5 \times 10^{51}$ ergcm^3, and for AD 185, $E_0/n \leqslant 10^{52}$ ergcm^3.

The above data on remnant age and diameter are plotted in figure 3.2, together with the relationship $D \simeq 0{\cdot}93t^{2/5}$, consistent with the statistical analysis of Clark and Caswell (1976), and the x-ray data discussed in Clark and Culhane (1976). It would appear that the remnants of the historical supernovae fit tolerably well the dynamical evolutionary track proposed from theoretical and statistical considerations.

Returning to the Crab Nebula, this has long been recognised as a remarkable object. The remnant, as currently observed at all wavelengths, clearly owes its existence to the continued injection of relativistic particles from its central pulsar, and many of the Crab Nebula's apparently unique properties may be attributed directly to the pulsar. However, it is of interest to consider how the expanding shock wave produced by the supernova outburst may have evolved, and whether its effects might be observable.

The shell of radio (and sometimes x-ray) emission seen in many SNR is usually assumed to delineate the position of the shock wave. For a remnant with the age of the Crab and at 2 kpc (the estimated distance of the Crab Nebula), the expected angular diameter of the shock wave would be about

24 arcmin, if $E_0/n \simeq 5 \times 10^{51}$ erg cm^3, typical of other SNR. The observed radio remnant has a diameter of approximately 5 arcmin and, if this were the diameter of the shock, it would imply a very low value of $E_0/n \simeq 2 \times 10^{48}$ erg cm^3. However, the possible existence of a weak radio shell with diameter approximately equal to 24 arcmin is not precluded by present measurements, since, as pointed out by Hazard and Sutton (1971), its detection would be difficult in the presence of the intense radio emission sustained by the pulsar. The bulk of the x-ray emission is confined to a source of small diameter, similar to the observed radio emission, but recent measurements (Toor *et al* 1976, Charles and Culhane 1977) show weak emission extending beyond the central region by several minutes of arc. Additional investigation is desirable to ascertain whether the emission extends further and, perhaps, traces out the shell for which we have been searching.

It is therefore possible that the unusual properties of the Crab Nebula are due to the central pulsar, and are *additional* to the more common shell remnant rather than in place of it.

It appears that there are a number of remnants within the Galaxy which show a general resemblance to the supernova of AD 1054 (Lockhart *et al* 1977). The most interesting of these is MSH 15–56, which shows not only an amorphous 'central' structure of unusually shallow radio spectrum, but also a well defined outer shell, presumably indicating the present position of the shock front (Clark *et al* 1975a, b).

Summarising the above discussion, apart from the Crab Nebula the historical supernovae individually suggest E_0/n values quite close to the mean value $E_0/n = 5 \times 10^{51}$ erg cm^3 derived by Clark and Caswell (1976). (The comparisons are, of course, fairly crude because they depend heavily on accurate measurements of distance d, $E_0/n \propto d^5$; so a 15% error in d results in a factor of two error in E_0/n.) Unless E_0 is proportional to n (which seems unlikely), the analysis implies that the range of values of both E_0 and n show quite low dispersions. Let us first consider the possible variations in n.

A realistic model of the interstellar medium should allow for:

(i) a diffuse intercloud gas of density n_{ic} (a value as small as

$0{\cdot}2\ cm^{-3}$ may be applicable) extending several hundred parsecs from the galactic plane, and

(ii) small-diameter ($\simeq$1 pc) clouds of greatly enhanced density, but showing the distribution of extreme population I type objects (i.e. localised to a few tens of parsecs from the plane).

Once the SNR is significantly larger than a typical cloud, the expanding shock wave should propagate through the diffuse intercloud medium without being affected significantly by the clouds (McKee and Cowie 1975). Indeed, Sgro (1975) has shown that a plane shock will completely recover its form within 2–3 cloud diameters of overtaking a cloud. In the Sedov equation (3.1), n_{ic} should thus be used. The dynamical evolution of remnants will then show, in accord with observation, little dependence on z distance up to several hundred parsecs — within which range most SNR are found. Thus for $n_{ic} \simeq 0{\cdot}2$, the corresponding typical value would be $E_0 = 10^{51}$ erg. From general considerations, it seems reasonable that, for such a catastrophic event as a supernova, the threshold might be well defined, with evolution beyond that threshold leading inevitably to supernova activity showing quite small dispersions in the explosion parameters.

The radio emission from SNR is of a synchrotron nature, although there is no completely acceptable theory for the origin of the required magnetic field and relativistic particles. It has been argued that the emission may, in fact, be dominated by quite distinct physical processes at various stages of its evolution. For 'old' remnants (of age greater than about 30 000 yr, and in a phase of their evolution where the adiabatic expansion relationship (3.1) no longer applies) the model of van der Laan (1962), in which the interstellar magnetic field compressed by the shock front and cosmic rays produces the radio emission, may suffice. For 'young' remnants (up to a few hundred years old) the model of Gull (1973) is plausible. In this, Rayleigh–Taylor instabilities at the interface between the ejecta and interstellar medium will form a zone of unstable convective mixing, enclosing 'knots' of greatly enhanced magnetic field sufficient to account for the strong radio emission. After a time, this convection zone expands and dissipates, so that no middle-aged remnants (of age greater than 1000 yr) would be expected to be radiating by this mechanism alone. Thus, in a uniform interstellar medium, there is no obvious reason why 'middle-aged' SNR in the adiabatic phase of their evolution should be strong radio emitters. Nevertheless, Clark and Caswell (1976) showed that the radio brightness of at least certain SNR appears to decrease monotonically with time throughout the adiabatic phase described by equation (3.1) (at least down to a limiting value of brightness corresponding to an age of close to 10 000 yr).

McKee and Cowie (1975) suggest that the optical emission of SNR in the adiabatic expansion phase comes principally from interaction of the shock wave with the denser cloudlets of the interstellar medium, and this is also the mechanism usually invoked for the patchiness of x-ray emission. Clark and Caswell (1976) speculated that much of the radio emission may also arise from this interaction. After the shock has crossed a cloud, a Rayleigh–Taylor instability will occur (McKee and Cowie 1975). This leads to the possibility that, for SNR in the adiabatic phase, non-thermal radio emission may occur due to the turbulent amplification of the magnetic field and acceleration of relativistic particles by convective motions, in a manner similar to the mechanism suggested by Gull (1973) for young SNR. If, as suggested above, the dense clouds are concentrated to within a few tens of parsecs of the galactic plane, then SNR at large values of z may be subluminous. However, their rate of expansion, dependent on n_{ic}, may not differ significantly from that of SNR close to the galactic plane. Additionally, SNR with diameters of several tens of parsecs would be expected to be significantly brighter on their edge nearest the plane — several SNR support this suggestion.

The above explanation might account for the particularly subluminous remnants of AD 1006 and AD 1054 (in the latter case, if the emission still being excited by the central pulsar is ignored), both of which are at several hundred parsecs from the plane. Note that such subluminous remnants would be under-represented in a sample selected from radio remnant detections relative to an 'historical' sample (which will, in fact, favour high-latitude objects suffering less obscuration). It seems clear that at least some SNR at quite large values of z are not subluminous in their radio emission; in such cases, perhaps they occur in regions with more cloudlets than is common at such z values. Indeed, there may be a distinct bias whereby potential supernovae occur preferentially in such regions with quite high concentrations of dense clouds.

We will now use the historical supernova data to estimate the frequency of such events within our Galaxy.

The frequency of galactic supernovae is particularly difficult to determine, and recent advances in the interpretation of observational data from SNR have failed to resolve the uncertainties. A statistical analysis of their radio SNR catalogue by Clark and Caswell (1976) suggested a characteristic interval between supernova events leaving long-lived remnants of order 150 yr. Such a large interval is clearly at variance with the much smaller estimates for supernovae in external spiral galaxies (of order 20 yr (Tammann 1974)) and with the most recent values for the birth-rate of pulsars. For example, Taylor and Manchester (1977) claim that the pulsar birth-rate required to maintain their implied galactic distribution is at least one every 40 yr, and may be as low as one every 6 yr, depending on the reliability of pulsar distances and resulting spatial density estimates. While it is by no means certain that all supernova events produce pulsars, and it may be possible that pulsar formation could result from a stellar collapse which does not precipitate an explosive envelope ejection, a reconciliation of the grossly disparate pulsar and radio SNR estimates of time interval is desirable.

Figure 3.3 depicts a 'bird's-eye' view of the Galaxy, indicating the positions of those SNR which are believed to be in the adiabatic phase of their evolution, and for which some estimate of distance is possible. It is interesting to note that the remnants of the seven (labelled) supernovae observed during the past two millennia are limited to a sector of the galactic plane with an included angle of only 60°. (Because of the rather uncertain nature of the new star of AD 386, it has been excluded from this study.) Furthermore, all lie at some distance from the galactic plane (approximately 80, 30, 250, 200, 430, 150 and 1190 pc, when chronologically ordered).

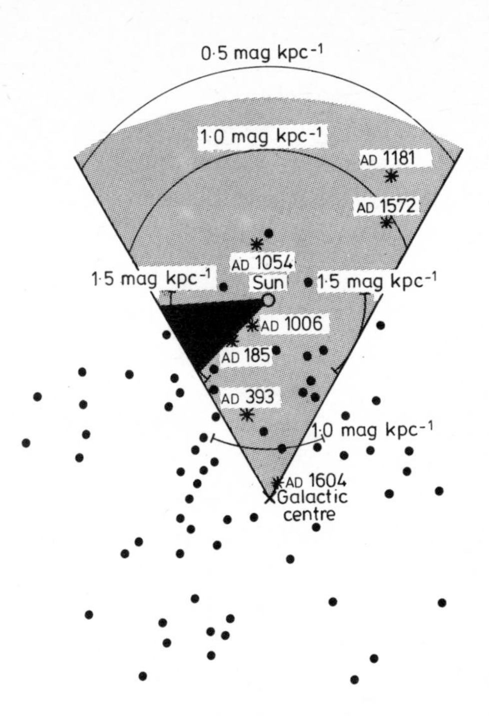

Figure 3.3. A 'bird's-eye' view of the Galaxy, showing the position of those radio remnants which are believed to be in the adiabatic phase of their evolution, and for which some estimate of distance is possible. The seven supernovae observed historically lie within the dotted 60° sector; the zone in heavy shading lies so far south that no recorded historical detection would be expected. Distance limits for the detection of supernovae are shown, assuming an apparent magnitude of +3 or brighter for a possible sighting and the indicated values of interstellar absorption.

This evidence suggests that supernovae lying outside the sector defined above, or lying closer than (say) about 50 pc from the plane, were unobservable because of obscuration by the dust that permeates the galactic disc.

We have previously argued that a new star is most unlikely to have been discovered with the unaided eye if fainter than about magnitude +3. If we assume that interstellar absorption near the galactic plane is equivalent to 1 mag kpc^{-1}, when, for example, looking across the spiral arms towards the

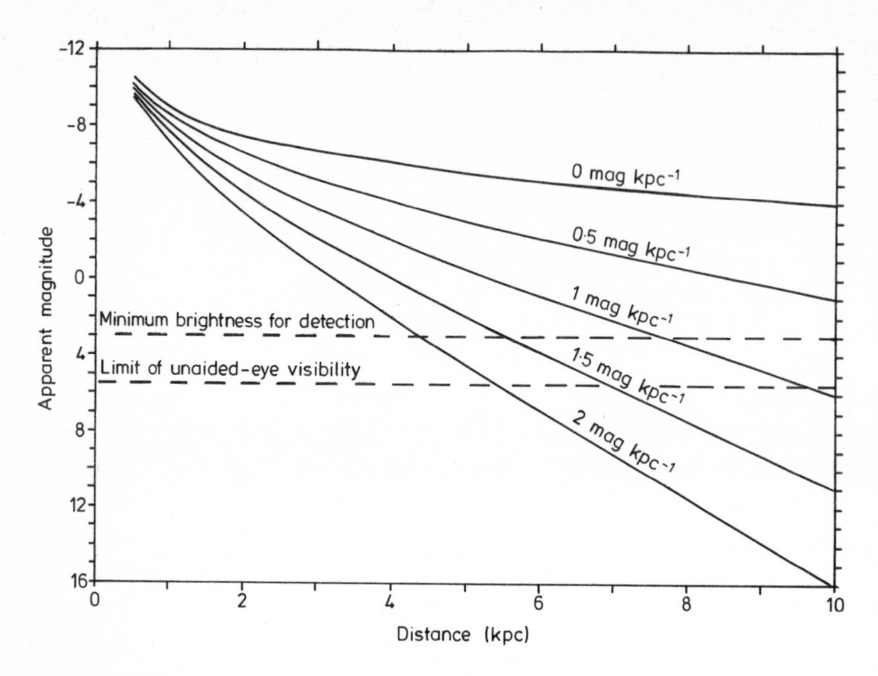

Figure 3.4. Curves of apparent magnitude of a supernova (absolute magnitude −19) as a function of distance for selected values of interstellar absorption between 0 and 2 mag kpc^{-1}.

galactic centre, then we can expect that a supernova will reach magnitude +3 only if it occurs at a distance of less than 7 kpc — see figure 3.4. (The detection of Kepler's supernova of AD 1604 can be explained by its extreme distance, over 1 kpc, from the galactic plane with consequent lower interstellar absorption.) In the anticentre direction, an expected lower absorption figure of (say) 0·5 to 1·0 mag kpc^{-1} would enable supernovae to be detected to the outer boundary of the Galaxy. Maximum obscuration, say 1·5 mag kpc^{-1}, would be expected in the galactic longitude ranges 50° to 90° and 260° to 320°, when looking nearly along the local spiral arm features. These directions can be recognised, for example, from the 'tangent points' in the galactic background radiation. The obscuration limits are indicated on figure 3.3 (assuming an apparent magnitude of +3 for the naked-eye detection of a supernova of absolute magnitude −19 — see figure 3.4), and would appear to explain successfully the observed surface distribution of the historical supernovae. This surface distribution suggests an incompleteness by at least a factor of *six* in the historical detections of galactic supernovae.

The galactic SNR tend to be concentrated within a few tens of parsecs of the galactic plane. Within 8 kpc of the galactic centre, their z distribution may be closely approximated by

$$N(z) = N(0) \exp(-|z|/60)$$

although beyond 8 kpc from the galactic centre, they are not so strongly concentrated towards the plane (Clark and Caswell 1976). As suggested earlier, this may not represent the true z distribution of galactic supernovae, but rather it reflects the suitability of the environment for the production of long-lived remnants, on the assumption that dense interstellar clouds near the plane play an important role in enhancing the radio brightness of 'middle-aged' SNR. Nevertheless, extragalactic studies do suggest that supernovae belong mainly to a flat, rather than a spherical, population. It is not possible to use the historical supernovae to deduce the z distribution of galactic supernovae, since historical detections would favour those events at large z value and with low obscuration. The scale height of young pulsars is 150 pc (Taylor and Manchester 1977). This is likely to be more indicative of that for supernovae than the SNR value of about 60 pc.

Additional incompleteness in the historical supernova sample could result from the non-detection of some objects which appeared near conjunction with the Sun (i.e. those lying beyond approximately 2 kpc and thus fainter than about apparent magnitude −4, and which have faded below magnitude +3 by the time of heliacal rising), and also from the fact that events in the southern skies (south of about declination −50°) would be unobservable from the northern hemisphere with its written histories.

It might also be thought that some historical records may have been lost, or some new star sightings never recorded. While this is undoubtedly the case for Europe and the Middle East, the completeness and essential reliability of the Far Eastern historical astronomical records is beyond dispute (Stephenson 1976, Stephenson and Clark 1977). The continuity of, for example, solar eclipse records from China makes it unlikely that selected astronomical records from the past two millennia have been lost — and the astrological significance of new star sightings makes it highly unlikely that any detection went unrecorded.

It is difficult to quantify accurately the effect of incompleteness in the historical sample that results from the non-detection of supernovae close to the plane, during daylight at distances greater than about 2 kpc, or in the southern skies. However, one might expect that collectively they could result in an incompleteness (within the 60° sector defined above) of an additional factor of 1·5 to 3.

The above argument suggests that in the past two millennia, a total of between 60 and 120 supernovae will have occurred. An average interval between supernovae within our Galaxy of 30 yr *or less* thus seems compatible with the available historical data. It also agrees well with growing evidence of such small intervals from surveys of extragalactic supernovae and with current estimates for the birth-rate of pulsars. (Tammann (1977) has considered independently the incompleteness of the historical data, to obtain a mean interval between galactic supernovae of 11^{+14}_{-4} yr. This may be compared with his value, implied from extragalactic evidence, of 20 yr, with an uncertainty of a factor two.)

Clark and Caswell (1976) have argued that their catalogue of radio SNR in the region covered by the Molonglo–Parkes survey (Clark *et al* 1973, 1975a, b, Caswell *et al* 1975, Caswell and Clark 1975) is essentially complete down to a surface brightness limit corresponding to an SNR age of greater than 10 000 yr. Their cumulative distribution for the number N of galactic SNR with diameter less than D (pc), estimated from the catalogued data, is

$$N(<D) \simeq 8 \times 10^{-3} D^{2\cdot5}. \quad (3.3)$$

Suppose only a fraction x of all supernovae leave long-lived radio remnants, so that $\tau = x\tau'$, where τ is the true mean interval between galactic supernovae and τ' is the characteristic interval between events leaving long-lived catalogued remnants. Then, assuming a constant rate of supernovae $N(D) = xt/\tau$, where t (yr) is the time taken for a remnant to expand to diameter D, from equation (3.3) we obtain

$$D = \left(\frac{xt10^3}{8\tau}\right)^{2/5}. \tag{3.4}$$

This corresponds to the well known Sedov formula for the adiabatic expansion of an SNR (equation (3.1)) when

$$x = 34{\cdot}4 \times 10^{-14}\tau(E_0/n)^{1/5}. \tag{3.5}$$

Taking $E_0/n \simeq 5\times10^{51}$ erg cm^3, and the value of $\tau \leqslant 30$ yr determined earlier, gives from equation (3.5)

$$x \leqslant 0{\cdot}22.$$

Thus the reconciliation of the available data on radio SNR (assuming the near completeness of existing catalogues) and the estimates of the mean interval between supernovae found from the historical sample and extragalactic data, requires that fewer than about one in five galactic supernovae leave long-lived radio remnants, although a much larger fraction must produce pulsars.

Presumably all supernovae will produce remnants which will initially brighten in the radio region — in accordance with the proposed mechanism of Gull (1973). Most will subsequently fade from view after approximately 1000 yr; possibly reappearing about 30 000 yr later with van der Laan (1962) type emission, but being difficult to identify with certainty because of their size as they merge with nearby objects and the galactic background radio emission. Only those 20% or so of supernovae occurring in a suitable environment, containing the dense cloudlets which it is believed may contribute significantly to the radio brightness of 'middle-aged' SNR, would produce remnants which would be detectable throughout their lifetime.

It is possible that galactic supernovae, with $E_0 \simeq 10^{51}$ erg, and occurring at intervals of 30 yr or less, could:

(i) provide most of the energy to maintain the random motions of interstellar clouds (Woltjer 1972);
(ii) explain the galactic distribution and birth-rate of pulsars;
(iii) be capable of producing the observed cosmic-ray flux (Scott and Chevalier 1975);
(iv) occur close enough to the Earth (say within 10 pc)

frequently enough (about every 10^8 yr) to have affected the Earth's biological and climatic history (Clark *et al* 1977a);

(v) sustain an interlinking SNR tunnel system, as proposed by Jones (1975); and

(vi) contribute significantly to the galactic x-ray and radio background.

The study of historically recorded new stars appears to have reached maturity. Little progress in the discovery of historical records of additional early new stars seems likely, although it is possible that further information on the stars we have discussed may yet appear. It is most unlikely that any new associations with SNR will be proposed, since catalogues of young SNR must now be considered complete.

The Crab Nebula is undoubtedly the best known remnant of an historical supernova. It may, however, have dominated the attention of astronomers for too long. It is to be hoped that the remnants of the other historical supernovae we have discussed may, in future, attract some of the observational attention so lavishly afforded the Crab Nebula. A unified study of the historical SNR must considerably advance our understanding of the nature of supernova explosions and the evolution of their remnants.

4. *Sunspots and Aurorae*

4.1. Introduction

Post-telescopic high-resolution observations at a variety of wavelengths have clearly established that, contrary to the historical view of solar perfection, the Sun does not radiate energy uniformly from its surface. Localised enhancements in energy emission are evidenced by such phenomena as photospheric dark sunspots beneath plages (brighter than average areas) in the chromosphere, flares (the explosive release of high-energy radiation and particles) producing terrestrial aurorae, prominences and non-uniform structures in the outer solar atmosphere (the corona). Of these various monitors of solar activity, the most easily observed and the only ones regularly recorded prior to the telescopic era were sunspots and aurorae. (In chapter 1, the only known unambiguous pre-telescopic detection of the solar corona was mentioned. There is a possible allusion to a solar flare on the An-yang oracle bones (Needham 1959) but this is of doubtful reliability.)

The mechanisms for the formation, maintenance and dissipation of solar activity are still only poorly understood, and these problems will only be resolved when the long-term solar behaviour is clearly established. In this chapter we look at what information on the past history of the Sun can be obtained from the historical sunspot and auroral records. In doing so, we recognise that sunspots themselves are not directly the cause of any of the established variations in solar activity. They are, however, a convenient index of almost all other activity on the Sun, and represent the only available source of solar data prior to the establishment of solar physics with the advent of the telescope.

4.2. Sunspot Records

A quasi-regular 11 yr cycle of solar activity has long been evidenced in plots of the annual means of observed sunspots, having first been proposed by Schwabe (1843) from 17 years of his observations. Wolf (1856) concluded that the cycle had been present since at least 1700, and subsequent investigations have shown it to be an enduring feature. Figure 4.1 shows the annual mean sunspot number between AD 1600 and 1975 (Eddy 1977). It should be emphasised that the statistical basis of the plot prior to 1850 is grossly inferior to

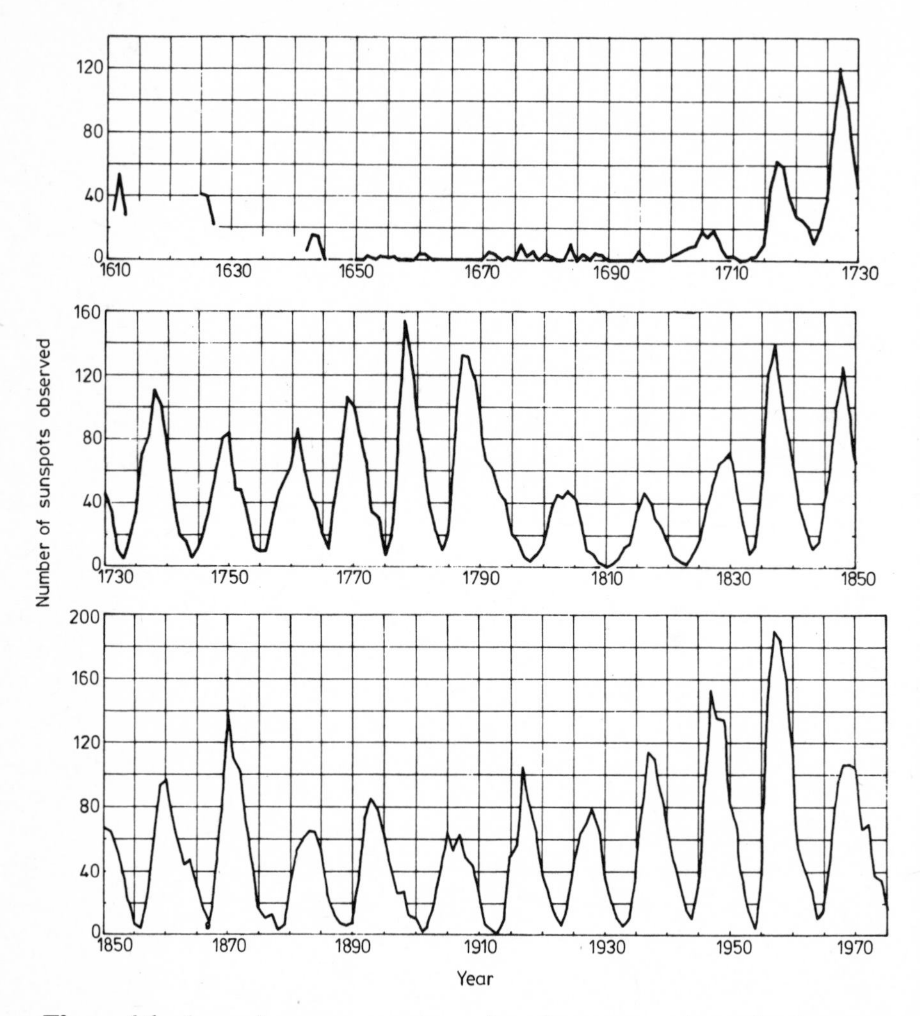

Figure 4.1. Annual mean sunspot numbers from post-telescopic observations (from Eddy 1976).

that following it. It is evident that the solar cycle 'period' over the past three centuries has varied between extremes of 8 and 15 yr (almost a factor of two), and that the cycle amplitudes have shown even wider variation.

Attention has recently been redrawn to an apparent protracted minimum in sunspot activity in the late seventeenth century, first reported by Spörer (1887, 1889) and subsequently by Maunder (1890). The period is now known as the Maunder Minimum, the rightful claim for its original discovery by Spörer apparently having been overlooked. Eddy (1976) has strengthened the reality of the claims that the sunspot cycle essentially vanished between 1645 and 1715 from a re-analysis of the contemporary literature detailing auroral records, sunspot counts and observations of the Sun at eclipse, plus indirect evidence from atmospheric isotopes. Indeed, Eddy claims that there is insufficient evidence to establish whether the 11 yr cycle existed before the onset of the Maunder Minimum and after the introduction of the telescope (although periods of significant sunspot activity were recorded). From direct telescopic records, the earliest clear evidence we have for the cyclic variation of solar activity is from about 1700.

There have been numerous attempts to extrapolate the sunspot cycle backwards in time, notably by Schove (1947, 1955, 1961, 1962), and most recently by Hill (1977). Eddy (1977) has questioned the validity of the techniques used to backdate solar cyclic behaviour, since they are based on the assumption that there has always been about nine cycles per century, whether or not there is observational evidence for them. Detailed spectral techniques using recent telescopic data to establish past behaviour (e.g. Hill 1977) fail to reproduce such gross features as the Maunder Minimum, and are clearly inadequate (see figure 4.2). The recent convincing evidence of several major excursions in solar activity during the past two millennia (Eddy 1977) necessitates, as a matter of urgency, a re-analysis of the historical data relating to past solar behaviour.

The Aristotelian concept of celestial perfection had become so rigidly accepted as a part of Christian theology through the middle ages that Galileo's questioning of the

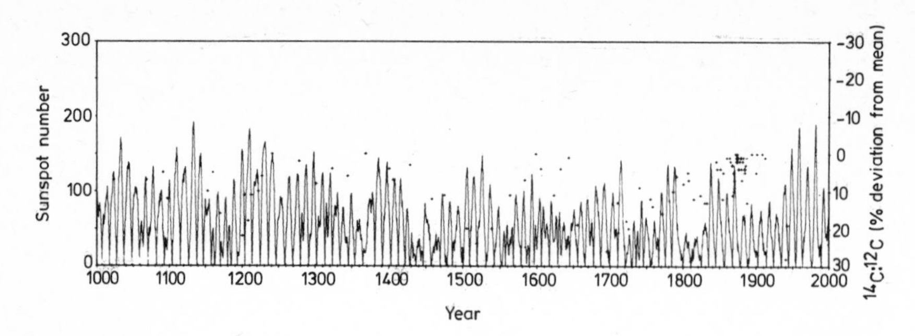

Figure 4.2. Fourier series sunspot model (full curve) and $^{14}C:^{12}C$ ratio (points, using an inverted scale) (from Hill 1977).

concept of solar faultlessness, following his telescopic 'discovery' of sunspots, was one of the reasons for his inquisition. In fact, the earliest known reference to a sunspot in the mid-fourth century BC can be attributed to a pupil of Aristotle, Theophrastus of Athens. Numerous subsequent sightings in the occident were incorrectly attributed to transits of Mercury and Venus, or 'Solar Moons'. The two instances of fourteenth-century sightings in Russia, mentioned in chapter 1, were at times of forest fires, emphasising the need for diminished brightness by atmospheric haze, dust storms, severe atmospheric absorption (near sunrise or sunset), reflection off still waters, etc, to render sunspots visible to the naked eye. Cases of the recording of such events as actual solar imperfections prior to the seventeenth century (e.g. in Russia) suggest a lack of acquaintance with Aristotelian doctrine.

The Orient was free from the stranglehold of this theory of celestial perfection, and sunspots were recorded from very early times. The first definite recording dates from 28 BC, and subsequent sightings are listed as *hei-ch'i* ('black vapour'), *hei-tzu* ('black spot') or *wu*. *Wu* means 'crow' as well as 'black', and Needham (1959) raises the possibility that the 'crow in the Sun' (the colleague of the 'rabbit on the Moon'), which was part of traditional Chinese mythology predating 28 BC, might be based on sunspot observations in legendary times.

The abundance of historical sunspot records from China may result not only from the open acceptance of celestial variability but also from the regular occurrence of conditions suitable for their detection. The geology of China renders it susceptible to dust storms, particularly off the Gobi and Tarim plains. In addition, there is a high incidence of atmospheric haze associated with a continental climate. Many of the Chinese sunspot records mention reduced solar brightness commensurate with these conditions: for example, 'within the Sun there was a black spot; the Sun had no brilliance' (AD 1145); 'the Sun was orange in colour . . .' (AD 189). Needham (1959) has suggested the possibility that the Sun might have been viewed through pieces of semi-transparent jade, mica, or smoky rock crystal, although we have been unable to find any confirmation of this suggestion in the available records.

As discussed in chapter 1, the oriental descriptions of sunspots are particularly picturesque, often attempting some estimate of size or shape. Sunspots are described as 'like a plum', 'as large as a date', 'like a hen's egg', 'as large as a peach', 'like a man', etc. One can only speculate on the true sunspot size from these classification criteria. Several simultaneously resolvable spots are often mentioned (e.g. 'within the Sun there were black spots, sometimes two, sometimes three, as large as chestnuts' — AD 1112 May 2), as well as large groups (see the example in chapter 1). The duration of visibility is often given (e.g. 'Within the Sun there was a black spot as large as a plum for a decade of days. It finally melted away.' — AD 1137). A particularly fascinating account dates from AD 189: 'The Sun was orange in colour and within it there was a black vapour like a flying magpie. After several *months* it melted away.' This sunspot group was apparently tracked for several solar rotations, although no mention of actual disappearance and reappearance is included.

In table 4.1 we have listed all the pre-telescopic sunspot observations we have found from oriental dynastic histories, diaries, etc, and these data are depicted graphically in figure 4.3. Not all of these are independent events: the same sunspot activity sometimes being sighted and recorded in China and Korea. (The very prominent peak near AD 1370, for example, results partly from dual recordings.) It is apparent that the occurrence of sunspot records is highly fragmentary and sporadic. Both Eddy (1977) and Bray (1974) have

Table 4.1. Pre-telescopic sunspot records from the Orient.

Julian date	Place	No.	Duration	Description
BC 28 May 10 (?)	C	1	—	As large as a coin
AD 20 Mar 17	C	1	—	—
187 Mar–Apr (?)	C	1?	—	As large as a melon
188 Feb–Mar (?)	C	1?	Several months	Like a flying magpie
268 Nov–Dec (?)	C	1	—	—
299 Feb–Mar (?)	C	1?	Several days	Like a flying swallow
301 Jan 19	C	1	—	—
301 Oct 20	C	1	—	—
302 Dec–Jan (?)	C	1	—	—
304 Dec–Jan	C	1	—	—
311 Apr 7	C	1?	—	Like a flying swallow
321 May 7	C	1	—	—
322 Nov 6	C	1	—	—
342 Mar 7	C	1	4 days	—
352–353	C	1?	5 days	The shape of a three-legged crow
354 Nov 7	C	1	—	As large as a hen's egg
355 Apr 4	C	2	—	As large as peaches
359 Nov 7	C	1	—	As large as a hen's egg
369 Nov 27 (?)	C	1	—	—
370 Mar 29	C	1	—	As large as a pear
373 Jan 28	C	1	—	—
373 Dec 26	C	1	—	As large as a pear
374 Apr 6	C	2	—	As large as duck's eggs
375 Jan 10	C	1	—	As large as a hen's egg
388 Apr 2	C	2	—	As large as pears
389 Jul 17	C	1	—	As large as a pear
395 Dec 13	C	1	—	—
400 Dec 6	C	1	—	—
499 Jul 4 (?)	C	1	—	—
500 Jan 29	C	1	—	As large as a peach
500 Jan 30	C	3	—	—
501 Sep 4	C	1	—	—
502 Feb 8	C	1	4 days (?)	Like a goose's egg
502 Feb 12 (?)	C	2	—	—
502 Mar 26 (?)	C	1	—	As large as a goose's egg

Table 4.1. (*continued*)

Julian date	Place	No.	Duration	Description
AD 510 Mar 17	C	2	—	—
511 Dec 16	C	2	—	As large as peaches
513 Apr 7	C	1	—	—
566 Mar 29	C	1	—	A crow was seen
567 Dec 10	C	1	9 days	As large as a cup
567 Dec 13	C	1	6 days	—
578 Dec 25	C	1	—	As large as a cup
579 Apr 3	C	1	4 days	As large as a hen's egg
826 May 7	C	1	—	Like a cup
826 May 24	C	1	—	—
832 Apr 21	C	1	—	—
837 Dec 22	C	1	—	As large as a hen's egg
841 Dec 30	C	1	—	—
851 Dec 22	J	1	—	As large as a plum
865 Feb–Mar	C	1	—	Like a hen's egg
874–875 (?)	C	1	—	—
875–876 (?)	C	1?	—	Like a flying swallow
904 Feb 19	C	7?	—	The Northern Dipper was seen within the Sun
927 Mar 9	C	1	—	Like a hen's egg in shape
974 Mar 3 (?)	C	2	—	—
1077 Mar 7	C	1	14 days	Like a plum
1077 Jun 7	NC	1	—	—
1078 Mar 11	C	1	19 days	Like a plum
1079 Jan 11	C	1	12 days	—
1079 Mar 20	C	1	9 days	As large as a plum
1104 Dec 11 (?)	C	1	—	As large as a date
1112 May 2	C	2–3	—	As large as chestnuts
1118 Dec 18	C	1	—	As large as a plum
1120 Jun 7	C	1	—	As large as a date
1122 Jan 10	C	1	—	As large as a plum
1129 Mar 22	C, NC	1	23 days	—
1131 Mar 12	C	1	4 days	—
1136 Nov 23	C	1	4 days	As large as a plum
1136 Nov 27	NC	2?	—	—
1137 Mar 1	C	1	10 days	As large as a plum
1137 May 8	C	1	14 (?) days	—

Table 4.1 (*continued*)

Julian date	Place	No.	Duration	Description
AD 1138 Mar 16	C	1	—	—
1138 Nov 26	C	1	—	—
1139 Mar	C	1	More than a month	—
1139 Nov 20	C	1	—	—
1145 Jul 23	C	1	—	—
1151 Mar 21	K	1	—	As large as a hen's egg
1160 Feb 28	K	1	3 days	—
1160 Sep 26	NC	1	—	Like a man in shape
1160 Sep 29	K	1	—	—
1171 Oct 20	K	1	—	As large as a peach
1171 Nov 16	K	1	—	As large as a peach
1183 Dec 4	K	1	2 days	—
1185 Feb 10	C	1	—	As large as a date
1185 Feb 11	K	1	—	As large as a pear
1185 Feb 15	C	1	12 days	—
1185 Mar 27	K	1	—	As large as a pear
1185 Apr 18	K	1	2 days	—
1185 Nov 14	K	1	—	—
1186 May 23	C	1	—	As large as a date
1186 May 26	C	1	—	—
1193 Dec 3	C	1	9 days	—
1200 Sep 19	K	1	—	As large as a plum
1200 Sep 21	C	1	5 days	As large as a date
1201 Jan 9	C	1	20 days	—
1201 Apr 6	K	1	—	As large as a plum
1202 Aug 23	K	1	—	As large as a pear
1202 Dec 19	C	1	12 days	As large as a date
1204 Feb 3	K	1	3 days	As large as a plum
1204 Feb 21	C	1	—	As large as a date
1205 May 4	C	1	—	—
1238 Dec 5	C	1	—	—
1258 Sep 15	K	1	—	As large as a hen's egg
1258 Sep 16	K	1?	—	Like a man in shape
1276 Feb 17	C	2?	—	Like goose's eggs
1278 Aug 31	K	1	—	As large as a hen's egg

Table 4.1 (*continued*)

Julian date	Place	No.	Duration	Description
AD1356 Apr 4	K	1	2 days	
1361 Mar 16	K	1	—	—
1362 Oct 5	K	1	—	—
1370 Jan 1	C	1	—	—
1370 Oct 2	C	1	—	—
1370 Oct 21	C	1	—	—
1370 Dec 7	C	1	—	—
1370 Dec–Jan	C	Several	1 month	Seen frequently
1371 Mar 31	C	1	—	—
1371 Jun 14	C	1	—	—
1371 Nov 6	C	1	—	—
1371 Nov 21	K	1	—	—
1372 Feb 6	C	1	—	—
1372 Apr 3	C	1	—	—
1372 May 8	K	1	—	—
1372 Jun 19	C	1	—	—
1372 Aug 25	C	1	—	—
1373 Apr 26	K	1	2 days	—
1373 Oct 23	K	1	—	—
1373 Nov 15	C	1	—	—
1374 Mar 27	C	1	4 days	—
1375 Mar 20	K	1	2 days	—
1375 Mar 23	C	1	—	—
1375 Oct 21	C	1	—	—
1376 Jan 19	C	1	—	—
1381 Mar 22	C	1	3 days	—
1381 Mar 23	K	1	—	—
1382 Mar 9	K	1	3 days	As large as a hen's egg
1382 Mar 21	C	1	—	—
1383 Jan 10	C	1	—	—
1387 Apr 15	K	1	—	—
1402 Nov 15	K	1	—	—
1520 Mar 9	K	2?	—	—
1603 Apr 16	K	3	—	Like large cash in shape
1604 Oct 25	K	1	—	As large as a hen's egg

C, China; NC, Northern China; J, Japan; K, Korea.

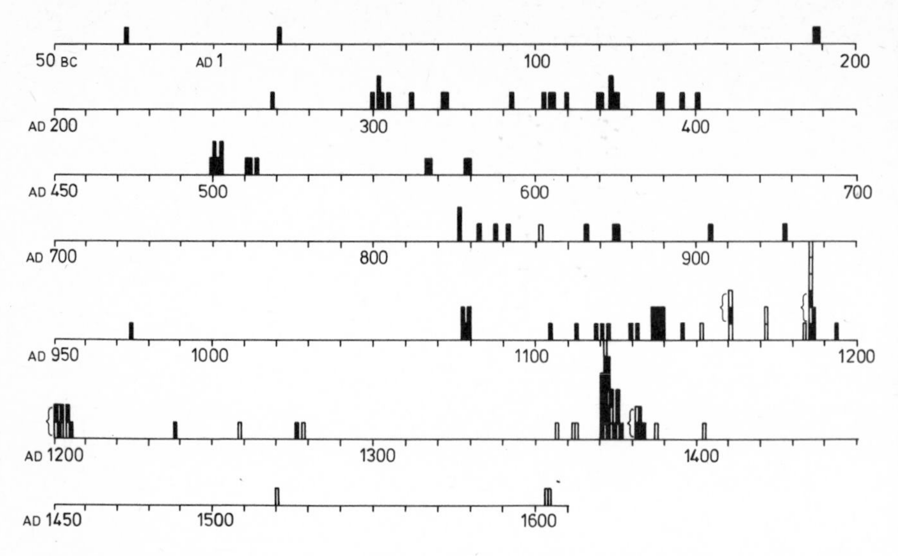

Figure 4.3. The pre-telescopic oriental sunspot sightings. Each 'bar' represents a single sighting; full bars from China, open bars from Korea or Japan.

suggested that the observations were apparently merely accidental, Bray claiming that their occurrences were gleaned from casual mention in the diaries of the *literati*. We cannot agree with this interpretation. It is true that no well documented evidence exists that a regular watch (perhaps at sunset) was kept for sunspots throughout oriental history. But, at least during certain periods, the acknowledged systematic and careful approach to observational astronomy and its compendious documentation, the abundance of records of various other daytime phenomena, and the deeply held belief that celestial phenomena were precursors of terrestrial events (all discussed in chapter 1) make it unlikely that only sunspots were omitted from regular patrol.

Clearly it is important to check the above interpretation before trying to search for periodicities in the historical sunspot records. We have attempted to do this by comparing the frequency of sunspot records for a particular dynasty with the frequency of records of other celestial phenomena, and also by comparing sunspot records from both China and Korea for a particular period. This study reveals a number of difficulties inherent in an interpretation of the historical sunspot data.

Figure 4.4 illustrates the statistics of selected astronomical observations made during the Chin dynasty in China. The completeness of the solar eclipse records results from their importance for calendrical purposes (with astrology, one of the main functions of the Astronomical Bureau). It emphasises that any incompleteness in the records of other phenomena did not result from the total loss, or destruction, of the astronomical records, or their total exclusion from a dynastic history for the period of a particular reign. However, there are certain enigmatic gaps in the records of other phenomena, particularly in all the records prior to AD 300, and in the sunspot records after AD 400. The almost total lack of astronomical records except for eclipses and occasional conjunctions prior to AD 300 calls into question the completeness of the documentation of celestial events of astrological significance at this time. By contrast, the lack of any sunspot records between AD 400 and 420 probably represents a total lack of any sightings, since here there is an abundance of records of all other phenomena.

Bielenstein (1950) has suggested that, while, with very few exceptions, records were not actually falsified, they were often left incomplete — the degree of incompleteness being related to the popularity of a particular reign. If 'warnings from Heaven' seemed not to be required, they might not be memorialised. Certainly, there is substantial evidence to support Bielenstein's view that 'popularity' meant popularity with the high officials of court who dictated what should be recorded by the historiographers, rather than with the mass of the people. An abundance of records of astrological significance with their accompanying admonitions and prognostications are often found to precede the overthrow of a dynasty (see, for example, figure 4.4).

Figure 4.5 summarises sunspot observations between AD 900 and 1400 taken from the histories of the Sung and Yüan dynasties in China, and also from the Korean *Koryŏ-sa* ('History of the Kingdom of Koryŏ' — AD 932–1392). The Sung Dynasty astronomers were particularly noted for their diligence and the accuracy of their astronomical descriptions — as evidenced by their records of the supernovae of AD 1006, 1054 and 1181 (Clark and Stephenson

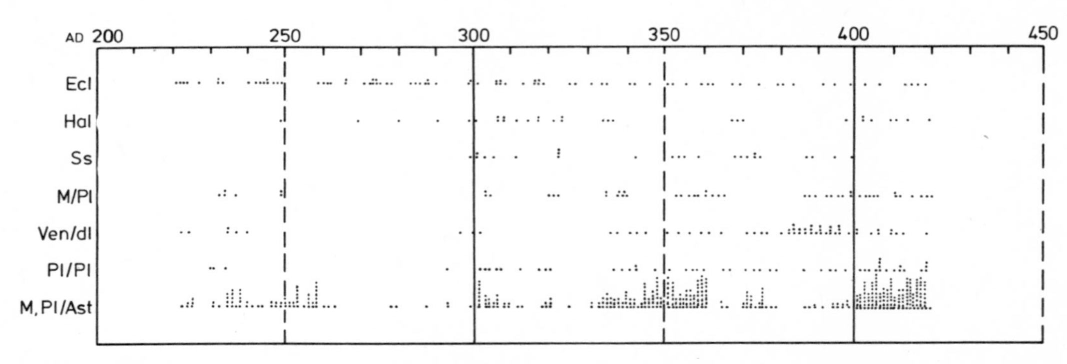

Figure 4.4. The statistics of various astronomical observations from the Chin dynasty.. Ecl, solar eclipses; Hal, solar haloes, etc; Ss, sunspots; M/Pl, occultations and conjunctions involving the Moon and a planet; Ven/dl, sightings of Venus in daylight; Pl/Pl, conjunctions of two or more planets; M, Pl/Ast, conjunctions of the Moon or a planet with an asterism or star.

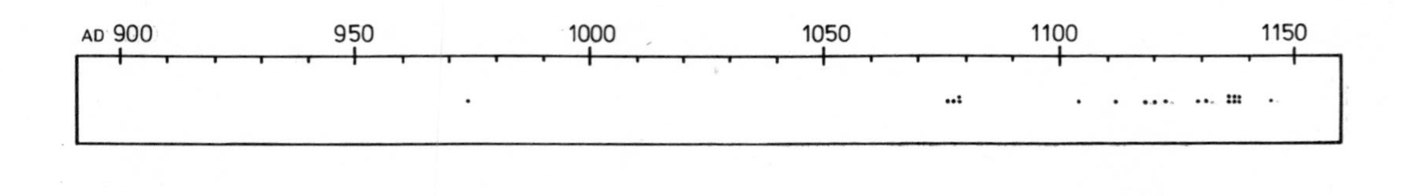

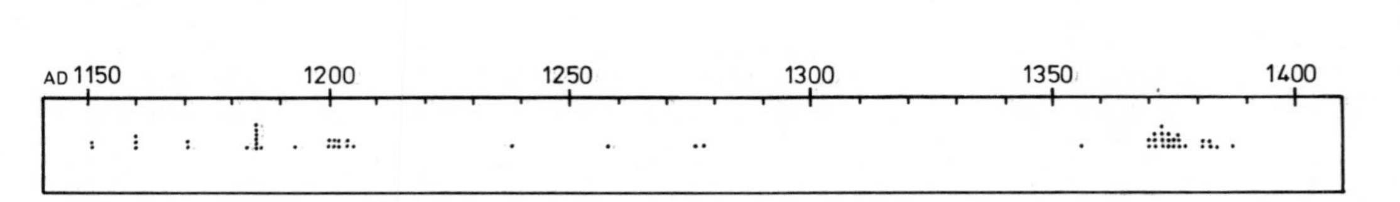

Figure 4.5. Sunspot observations from the Sung and Yüan dynastic histories, and the Korean *Koryŏ-sa*.

1977). The abundance of sunspot records from China between AD 1050 and 1250 is therefore hardly surprising, and this interest was maintained through the Yüan dynasty. What is surprising is the sudden onset of interest in Korea, emphasised in figure 4.6. The similarity in style of descriptions of sunspots suggests that the Korean interest had its origins in China.

After the amalgamation of the Three Kingdoms in AD 936 into the single kingdom of Koryŏ, almost a century was to pass before the establishment of an astronomical/astrological system based firmly on the Chinese model. Detailed records of a variety of phenomena date from about AD 1010. But it is not until AD 1150 that the first sunspot is recorded, followed by 60 yr of the most dramatic series of pre-telescopic sunspot records in existence. Between AD 1150 and 1210, we have the clearest available evidence of 'periodicity' in the grouping of any historical sunspot data. There can be little doubt that, at least during these times in Korea, a regular watch was being kept of the Sun for sunspot activity.

It is only the largest of individual spots, or unresolved large sunspot groups, that can be detected easily with the naked eye. Thus, since such sunspot configurations are now recognised as being most common near the maximum of the sunspot cycle, historical naked-eye sightings would be expected to delineate times of enhanced solar activity. (Note, however, that isolated large spots can be sighted throughout the sunspot cycle.) Figure 4.6 thus suggests the years AD 1151, 1160, 1171, 1185 and 1202 . . . 1355, 1362, 1373 and 1382 as lying near (±2 yr) epochs of maxima in past solar cycle 'on-states' (however transient these may have been). These years are nearly coincident with the reconstruction by Schove (1955), based on the historical sunspot records as well as other indicators. However, they *do not* coincide with the maxima in the sunspot model of Hill (1977), being closer to minima, again emphasising the inadequacies of the model and the difficulties in extrapolating solar behaviour into pre-telescopic times from a spectral analysis of telescopic data.

It is evident that, apart from periods when sunspots were clearly of great interest and presumably searched for regularly, the sporadic nature of the historical sunspot records

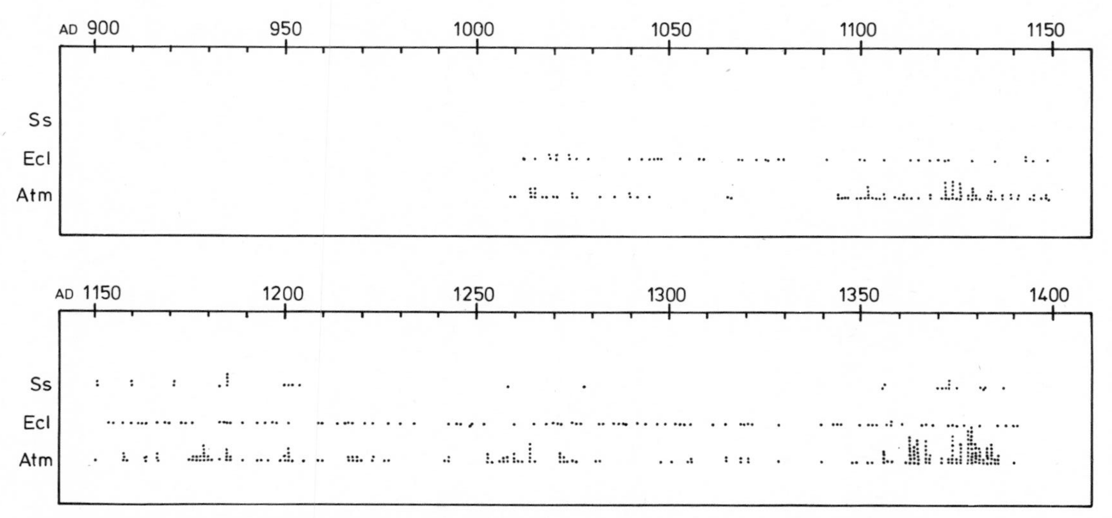

Figure 4.6. Statistics of astronomical observations from Korea. Ss, sunspots; Ecl, solar eclipses; Atm, atmospheric phenomena.

and their limited number means that they are, alone, rather poor indicators of past solar behaviour.

Referring again to figure 4.3, the two periods when many sunspots were recorded, in the fourth century and twelfth century, indicated times of intense interest in sunspots — during the Chin Dynasty in China, and then in Korea, respectively. As noted earlier, the prominent peak in records near AD 1370 results partly from dual recordings (in China and Korea). Periods of scarcity of records might result from protracted minima in sunspot activity, similar to the Maunder Minimum; equally, they may merely result from lack of interest. Further research is required to try to establish whether sunspots were observed, or not, at these times — and if they were, why they were not memorialised.

4.3. Auroral Records

The sunspot data can be supplemented with historical auroral observations. As we shall see later, the composite data can, at best, merely be used as a very coarse check on long-term solar variability monitored by indirect methods.

Aurorae are produced by the precipitation into a narrow band of the atmosphere (approximate latitudes 65° to 75°) of energetic particles, originating in the solar wind and entering the Earth's magnetic field via high-latitude clefts on the day side and the magnetospheric tail on the night side. The occurrence of aurorae is thus related directly to the occurrence of solar flares and enhancements in the solar wind flux, and aurorae are, as a consequence, a crude indicator of solar activity.

Fritz (1873) prepared a catalogue of historical observations of aurorae, principally from Europe. Recognised as an atmospheric phenomenon, aurorae did not conflict with Aristotelian dogma, and are historically recorded in abundance from Northern Europe. (There are about 50 accounts of comparatively rare mid-latitude aurorae from the Orient.)

We have reproduced as figure 4.7 a plot from Eddy (1977) summarising the data from Fritz's (1873) catalogue. The data are limited to reports from latitudes south of the Polar Circle, to exclude the regions near the magnetic poles where the

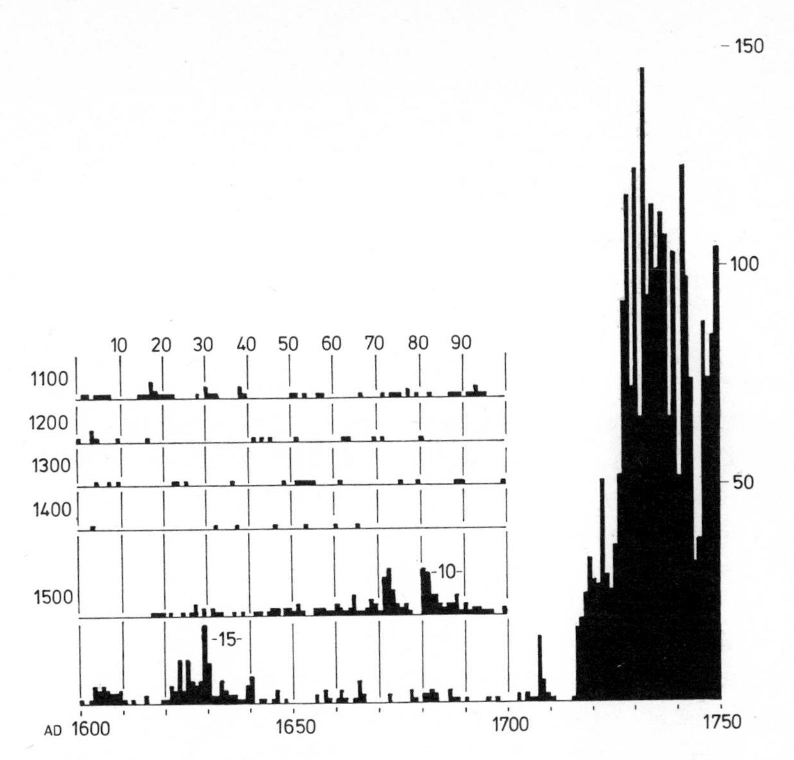

Figure 4.7. Reports of aurorae from the catalogue of Fritz (1873) (from Eddy 1977).

persistence of auroral phenomena smothers any solar-cycle dependence. The scarcity of records in historical times makes a search for solar-cycle periodicity fruitless.

The dramatic increase in the number of auroral records in the sixteenth century unquestionably reflects the release from intellectual prejudice and superstition at the time of the Renaissance. As pointed out by Eddy, the dramatic increase in reports from 1716 coincides with the publication of a classic paper on the aurora by Edmund Halley (1716), which inevitably influenced the number subsequently reported. What is probably a genuine reflection of actual solar behaviour is the dearth of records in the latter half of the seventeenth century — coinciding with the Maunder Minimum. Also of interest is the large number of auroral records in the twelfth century compared with those preceding and following. As noted earlier, this was the time of most prolific recording of sunspots in the Orient. Note, however, that there appears to be no peak in recorded auroral activity to correspond with that for sunspots near AD 1370–1380.

4.4. Other Indicators of Solar Activity

It has been suggested by several investigators in the past that natural 'unwritten' records, such as tree rings and ^{14}C data, may extend our knowledge of solar activity back well beyond the beginning of written history — and these data are obviously free from the severe selection effects and prejudices evident in the written records.

Unfortunately, tree-ring widths fail to show convincing evidence of past solar cyclic behaviour (La Marche and Fritts 1972). Local, rather than global, effects dominate the tree-ring patterns. In the absence of complicating influences, such as the crowding of other trees, etc, the widths of annual growth rings do monitor local rainfall during the growing season, and are therefore believed to be useful local climate indicators. Recent investigations have failed to produce convincing correlations between local climatic conditions and solar behaviour; thus, regretfully, there seems little future in pursuing the use of tree-ring widths in studies of long-term solar behaviour.

In contrast, the usefulness of ^{14}C data for studying the Sun's history has been proven convincingly (see, for example, Bray 1968, 1971). ^{14}C is produced in the upper atmosphere as the result of cosmic-ray bombardment. The extended solar magnetic field modulates the cosmic-ray flux at the Earth: during times of high solar activity the enhanced solar magnetic field reduces the cosmic-ray flux reaching the Earth and, consequently, the ^{14}C production (and conversely). Thus the relationship between ^{14}C production and sunspot number is an inverse one, at least in the short term (say the past few thousand years), when the effect of the varying intensity of the Earth's magnetic field can be accounted for.

A record of the ^{14}C content of the atmosphere throughout history is preserved in carbonaceous fossil material — for example, in trees. ^{14}C is assimilated as CO_2 in the photosynthesis process, and individual datable tree rings thus provide estimates of the ^{14}C:^{12}C abundance ratio. Such records exist in well preserved dead wood to beyond 5000 BC — and in living trees to about 3000 BC.

Figure 4.8 shows a compilation of ^{14}C data from Lin *et al* (1975), plotting the deviation (increased ^{14}C downwards) from an 1890 'norm' (shown as a broken line). The sinusoidal curve fitted by Lin *et al* matches well the smoothed curve of changing magnetic moment of the Earth, revealed from the interpretation of palaeomagnetic data, so that the overall envelope in the ^{14}C history clearly reflects the cyclic variations in the geomagnetic field. Major excursions from the smoothed curve would be expected to evidence extremes in solar behaviour, and, indeed, Eddy (1976) has shown that the Maunder Minimum corresponds to a pronounced ^{14}C deviation of about 10 parts per million.

It is impossible to recognise short-term fluctuations (such as the 11 yr cycle) in the ^{14}C history because of the appreciable delay (10–50 yr) between variations in ^{14}C production and resultant changes in the biospheric abundance. Nonetheless, Eddy (1977) has used the ^{14}C history to recognise six major excursions in solar behaviour in the past two millennia (see figure 4.9) with possibly a total of 12 in the past five millennia.

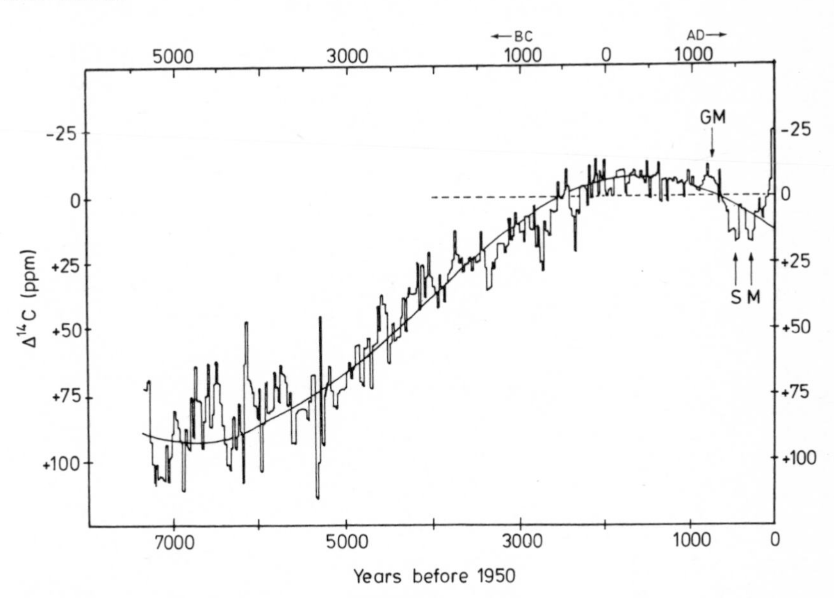

Figure 4.8. Deviations of ^{14}C concentrations from tree-ring analyses (in parts per million) (from Lin *et al* 1975).

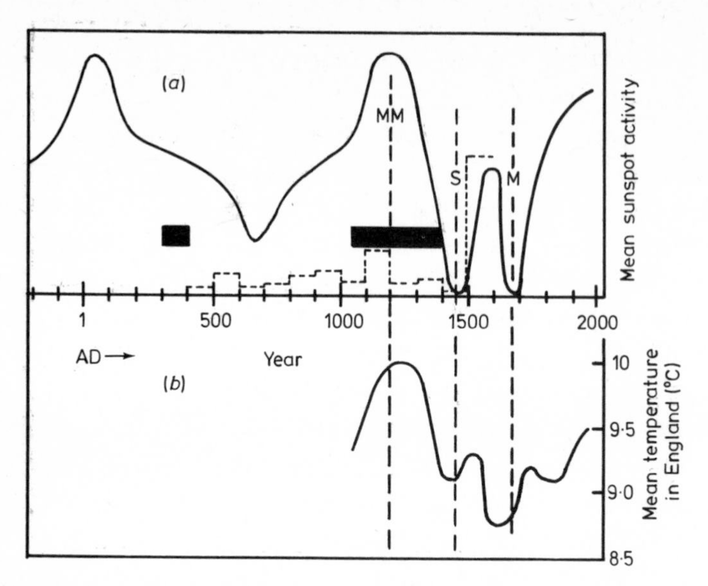

Figure 4.9. (*a*) The ^{14}C history as a guide to solar activity (from Eddy 1977). The periods of extensive historical sunspot sightings have been added as shaded horizontal bars, and the number of auroral records per century as a dotted histogram. (*b*) Estimates of mean annual temperature in England (from Lamb 1972).

In figure 4.9(*a*) the ^{14}C history is used as a guide to solar activity. The Maunder Minimum — a minimum in sunspot activity being represented by a positive (increased) deviation in ^{14}C (plotted downwards) — is labelled as ‘M’. A deviation of comparable magnitude in the fifteenth century, referred to by Eddy (1976) as the Spörer Minimum, is labelled with an ‘S’. Certainly during this time (1400–1500) there is a dearth of sunspot and auroral records. In figure 4.9(*b*) is plotted an estimate of the mean annual temperature in England over the past millennium, from Lamb (1972). The similarities with the curve in (*a*) are striking, suggesting that although attempted short-term correlations between solar activity and climate have proved to be inconclusive, long-term climatic variations are dominated by solar behaviour. The mode of this influence remains uncertain.

In addition to the Spörer and Maunder Minima, a less pronounced minimum (the ‘Mediaeval Minimum’) occurred between about AD 600 and 800. Between the Mediaeval and

Spörer Minima lies the twelfth-century 'Mediaeval Maximum' 'MM' (also referred to as the 'Grand Maximum') — a similar maximum (the so-called 'Roman Maximum') occurred during the first century, and the envelope would appear to be approaching a 'Modern Maximum'. Eddy (1977) emphasises that there is no periodic behaviour in these (and earlier) major excursions. He asserts that significant auroral activity, evidence of a structured solar corona, and extensive cyclic sunspot activity are not permanent features of the solar behaviour, but reveal themselves only at times of excursion maxima. As figure 4.9(*a*) indicates, such excursion maxima occur very infrequently, so that the above-mentioned solar activity indicators represent the exception to the normal solar condition in the Eddy model.

How do the historical sunspot and auroral records discussed earlier in this chapter fit such an interpretation?

In figure 4.9(*a*), the two extended periods of pre-telescopic oriental sunspot sightings are indicated, and also the total number of auroral records per century. Both the Spörer and Mediaeval Minima are characterised by an almost total lack of records of either phenomenon — by contrast, the Grand Maximum corresponds to the period of a wealth of sightings for both (plus evidence of 11 yr periodicity in the sunspot data). We have already stressed the social, political and astrological influences inherent in the pre-telescopic histories, so that, while the naked-eye sunspot and auroral records could not be used as conclusive evidence in support of the Eddy model, they are certainly not inconsistent with such an interpretation.

The suggested control of the sunspot cycle by planetary tidal effects has its origin in the near coincidence of the sidereal period of Jupiter (11·87 yr) and the average period of mean sunspot numbers (11·3 yr). The possible configurations of the other planets then allows a wide range of other tidal periods and effects. Detailed analyses have been completed by Wood and Wood (1965), and others, and the idea has recently been expounded enthusiastically by Gribbin and Plagemann (1975). Sadly, the theory is inconsistent with the evidence presented above for major excursions in sunspot activity. As pointed out by Smythe and Eddy (1977), any tidal theory rests on the evidence of continued sunspot periodicity — and we have already shown that this is clearly not the case. Indeed, Smythe and Eddy show that during the Maunder Minimum there are no striking dissimilarities in the incidence of various planetary conjunction patterns compared with other historical periods; and their more quantitative test, based on estimates of tidal potentials, supports this conclusion. The origin of the solar cycle — and the mode of its apparent disappearance and re-appearance — remains one of the mysteries of astronomy.

In chapters 2 and 3 of this monograph we discussed areas where the pre-telescopic astronomical data have been used to solve problems in modern astrophysics. Clearly, the material introduced in chapter 4 still requires additional research, and this is continuing. Many other areas where the pre-telescopic data may have applications, such as in cometary investigations, planetary orbital perturbation studies, the checking of historical chronology, etc, have been alluded to in passing. The uses of the pre-telescopic astronomical records in geophysics and astrophysics are now clearly established, and we are sure that this vast mine of information still has many surprises in store.

References

Adams J C 1853 *Phil. Trans. R. Soc.* **143** 397
Baade W 1943 *Astrophys. J.* **97** 119
—— 1945 *Astrophys. J.* **102** 309
Barbon R, Ciatti F and Rosino L 1974 *Supernovae and Supernova Remnants* ed C B Cosmovici (Dordrecht: Reidel) pp 99 and 115
Bielenstein H 1950 *Bull. Mus. Far East Antiq.* **22** 127
Bray J R 1968 *Nature* **220** 672
—— 1971 *Adv. Ecol. Res.* **7** 177
—— 1974 *Scientific, Historical and Political Essays in Honor of Dirk J Struik: Boston Studies in the Philosophy of Science* vol 15, eds R S Cohen *et al* (Dordrecht: Reidel) p 142
Brecher K, Lieber E and Lieber A E 1978 *Nature* **273** 728
Brosche P 1967 *IAU Inf. Bull. No. 192*
Brown E W 1919 *Tables of the Motion of the Moon* (New Haven: Yale University Press)
Caswell J L and Clark D H 1975 *Aust. J. Phys., Astrophys. Suppl.* **37** 57
Caswell J L, Clark D H and Crawford D F 1975 *Aust. J. Phys., Astrophys. Suppl.* **37** 39
Charles P A and Culhane J L 1977 *Astrophys. J.* **211** L23
Chu Sun-Il 1968 *J. Korean Astron. Soc.* **1** 29
Clark D H and Caswell J L 1976 *Mon. Not. R. Astron. Soc.* **174** 267
Clark D H, Caswell J L and Green A 1973 *Nature* **246** 28
—— 1975a *Aust. J. Phys., Astrophys. Suppl.* **37** 1
Clark D H and Culhane J L 1976 *Mon. Not. R. Astron. Soc.* **175** 573
Clark D H, Green A and Caswell J L 1975b *Aust. J. Phys., Astrophys. Suppl.* **37** 75
Clark D H, McCrea W H and Stephenson F R 1977a *Nature* **265** 318
Clark D H, Parkinson J H and Stephenson F R 1977b *Q. J. R. Astron. Soc.* **18** 443
Clark D H and Stephenson F R 1975 *Observatory* **95** 190
—— 1976 *Q. J. R. Astron. Soc.* **17** 290
—— 1977 *The Historical Supernovae* (Oxford: Pergamon)
Clemence G M 1948 *Astron. J.* **53** 169
Corcoran T H (transl) 1972 *Seneca: Naturales Quaestiones* vol 2 (London: Heinemann)
Delaunay C E 1859 *Mém. Acad. Sci., Paris* **48** 817

Downes D 1971 *Astron. J.* **76** 305
Duin R M and Strom R G 1975 *Astron. Astrophys.* **39** 33
Eddy J A 1976 *Science* **192** 1189
—— 1977 *The Solar Output and its Variation* (Boulder: Colorado Associated University Press)
Fotheringham J K 1920 *Mon. Not. R. Astron. Soc.* **81** 104
Fritz H 1873 *Verzeichniss Beobachter Polarlichter* (Wien: C Gerold's Sohn)
Gribbin J and Plagemann W 1975 *The Jupiter Effect* (London: Fontana)
Gull S F 1973 *Mon. Not. R. Astron. Soc.* **161** 47
—— 1975 *Mon. Not. R. Astron. Soc.* **171** 237
Halley E 1695 *Phil. Trans. R. Soc.* **19** 160
—— 1715 *Phil. Trans. R. Soc.* **29** 255
—— 1716 *Phil. Trans. R. Soc.* **29** 406
Hazard C and Sutton J 1971 *Astrophys. Lett.* **7** 179
Henning K and Wendker H J 1975 *Astron. Astrophys.* **44** 91
Hill E R 1967 *Aust. J. Phys.* **20** 297
Hill J R 1977 *Nature* **266** 151
Ho Peng Yoke 1962 *Vistas Astron.* **5** 127
Ho Peng Yoke, Paar T H and Parsons P W 1970 *Vistas Astron.* **13** 1
Hsi Tsê-Tsung and Po Shu-Jen 1965 *Acta Astron. Sin.* **13** 1 (transl: *NASA Tech. Trans.* TTF-388)
Ilovaisky S A and Lequeux J 1972 *Astron. Astrophys.* **18** 169
Jones E M 1975 *Astrophys. J.* **201** 377
Kanda Shigeru 1935 *Nihon Temmon Shiryō* (Tokyo: Kōsheisha)
Kant I 1754 *Wöchentliche Fragund Anzeigungs-Nachrichten* **23** and **24**
Kiang T 1971 *Mem. R. Astron. Soc.* **76** 27
van der Laan H 1962 *Mon. Not. R. Astron. Soc.* **124** 179
La Marche V C and Fritts H C 1972 *Tree-Ring Bulletin* **32** 21
Lamb H H 1972 *Climate: Present, Past and Future* vol 1 (London: Methuen)
Langdon S and Fotheringham J K 1928 *The Venus Tablets of Ammizaduga* (London: Oxford University Press)
de Laplace P S 1788 *Mém. Acad. Sci., Paris* **235**
Lin Y C, Fan C Y, Damon P E and Wallick E J 1975 *Proc. 14th Int. Cosmic Ray Conf., Munchen* vol 3, p 995
Lockhart I A, Goss W M, Caswell J L and McAdam W B 1977 *Mon. Not. R. Astron. Soc.* **179** 147
Lundmark K 1921 *Publ. Astron. Soc. Pacific* **33** 225
McKee C F and Cowie L L 1975 *Astrophys. J.* **195** 715
Martin C F and Van Flandern T C 1970 *Science* **168** 246
Maunder E W 1890 *Mon. Not. R. Astron. Soc.* **50** 251
Milne D K 1970 *Aust. J. Phys.* **23** 425
Morrison L V and Ward C G 1975 *Mon. Not. R. Astron. Soc.* **173** 183
Muller P M 1976 *PhD Thesis* University of Newcastle-upon-Tyne

Muller P M and Stephenson F R 1975 *Growth Rhythms and History of the Earth's Rotation* eds G D Rosenberg and S K Runcorn (London: Wiley) p 459
Needham J 1959 *Science and Civilisation in China* vol 3 (London: Cambridge University Press)
Newcomb S 1895 *Astron. Pap. Am. Ephemeris* **6**
Newton R R 1968 *Geophys. J. R. Astron. Soc.* **14** 505
—— 1970 *Ancient Astronomical Observation and the Acceleration of the Earth and Moon* (Baltimore: Johns Hopkins Press)
—— 1972a *Mem. R. Astron. Soc.* **76** 99
—— 1972b *Medieval Chronicles and the Rotation of the Earth* (Baltimore: Johns Hopkins Press)
Oesterwinter C and Cohen C J 1972 *Celestial Mech.* **5** 317
Parker R A and Dubberstein W H 1956 *Babylonian Chronology 626* BC–AD *75* (Providence, RI: Brown University Press)
Payne-Gaposchkin C 1957 *The Galactic Novae* (Amsterdam: North-Holland)
Sachs A J 1948 *J. Cuneiform Stud.* **2** 271
Sawyer J F A and Stephenson F R 1970 *Bull. Sch. Orient. Afr. Stud.* **33** 467
Schove D J 1947 *Terr. Magn. Atmos. Electr.* **52** 233
—— 1955 *J. Geophys. Res.* **60** 127
—— 1961 *J. Br. Astron. Assoc.* **71** 320
—— 1962 *J. Br. Astron. Assoc.* **72** 30
Schwabe H 1843 *Astron. Nachr.* **20** No. 295
Scott J S and Chevalier R A 1975 *Astrophys. J.* **197** L5
Sedov L I 1959 *Similarity and Dimensional Methods in Mechanics* (New York: Academic Press)
Sgro A G 1975 *Astrophys. J.* **197** 621
de Sitter W 1927 *Bull. Astron. Inst. Neth.* **4** 21
Smith C F (transl) 1919 *Thucydides* vol 1 (London: Heinemann)
Smythe C M and Eddy J A 1977 *Nature* **266** 434
Spencer-Jones H 1932 *Ann. Cape Obs.* **13** part 3
—— 1939 *Mon. Not. R. Astron. Soc.* **99** 541
Spörer F W G 1887 *Astron. Ges. Vierteljahrsschr. Lpz* **22** 323
—— 1889 *Bull. Astron.* **6** 60
Stephenson F R 1974 *Phil. Trans. R. Soc.* **A276** 118
—— 1976 *Q. J. R. Astron. Soc.* **17** 121
Stephenson F R and Clark D H 1977 *Q. J. R. Astron. Soc.* **18** 340
Stephenson F R, Clark D H and Crawford D F 1977 *Mon. Not. R. Astron. Soc.* **180** 567
Tammann G A 1974 *Supernovae and Supernova Remnants* ed C B Cosmovici (Dordrecht: Reidel) p 215
—— 1977 *Proc. 8th Texas Symp. on Relativistic Astrophysics* (New York: New York Academy of Sciences)
Taylor J H and Manchester R N 1977 *Astrophys. J.* **215** 885

TIES 1925 *Trans. Illum. Engng Soc.* **20** 565
Toomer G J 1974 *Arch. Hist. Exact Sci.* **14** 126
Toor A, Palmieri T M and Seward F D 1976 *Astrophys. J.* **207** 96
Tuckerman B 1964 *Mem. Am. Phil. Soc.* **59**
Van Flandern T C 1975 *Mon. Not. R. Astron. Soc.* **170** 333
Vyssotsky A N 1949 *Medd. Lunds Astron. Obs.* Historical Papers 22
Wolf R 1856 *Astron. Mitt. Zürich* **1** 8
Woltjer L 1972 *Ann. Rev. Astron. Astrophys.* **10** 129
Wood R M and Wood K D 1965 *Nature* **208** 129

Index